Springer Laboratory

Springer Laboratory Manuals in Polymer Science

Pasch, Trathnigg: HPLC of Polymers
ISBN: 3-540-61689-9 (hardcover)
ISBN: 3-540-65551-4 (softcover)

Mori, Barth: Size Exclusion Chromatography
ISBN: 3-540-65635-9

Pasch, Schrepp: MALDI-TOF Mass Spectrometry of Synthetic Polymers
ISBN: 3-540-44259-6

Kulicke, Clasen: Viscosimetry of Polymers and Polyelectrolytes
ISBN: 3-540-40760-X

Hatada, Kitayama: NMR Spectroscopy of Polymers
ISBN: 3-540-40220-9

Brummer, R.: Rheology Essentials of Cosmetics and Food Emulsions
ISBN: 3-540-25553-2

Mächtle, W.; Börger, L.: Analytical Ultracentrifugation of Polymers and Nanoparticles
ISBN: 3-540-23432-2

Rüdiger Brummer

Rheology Essentials of Cosmetic and Food Emulsions

184 Figures, 139 in color and 18 Tables

Springer

Rüdiger Brummer
Beiersdorf AG
Unnastr. 48
20245 Hamburg
Germany
e-mail: Ruediger.Brummer@Beiersdorf.com

Library of Congress Control Number: 2005930321

ISBN-10 3-540-25553-2 **Springer Berlin Heidelberg New York**
ISBN-13 978-3-540-25553-6 **Springer Berlin Heidelberg New York**

Springer is a part of Springer Science+Business Media
springeronline.com

Cover design: KünkelLopka GmbH, Heidelberg; *design&production*, Heidelberg, Germany
Typesetting and production: LE-TEX Jelonek, Schmidt & Vöckler GbR, Leipzig, Germany

2/3141 YL 5 4 3 2 1 0 - Printed on acid-free paper

Springer Laboratory Manuals in Polymer Science

About the Author

Rüdiger Brummer is a physicist at Beiersdorf AG with research interests in pressure-sensitive adhesives and cosmetic emulsions. He has published several professional papers and is the holder of patents. He completed his physics degree at the Christian Albrecht University in Kiel. Beginning in 1978 he worked as a scientist in the basic research laboratory of the Dr. Beyschlag Company in Heide.

In 1981 he moved to Phoenix AG in Hamburg, where he worked in a development laboratory for metal–rubber materials and started programming with finite elements. After several years he joined Beiersdorf AG, where he started with rheological measurements. In 1991 he became head of the Rheology and Thermal Analysis Department at Beiersdorf AG in Hamburg.

Rüdiger Brummer is a member of the German Rheology Society and the American Society of Rheology. He is active in the German DIN for viscosity and in the IUPAC sub-committee Structure and Properties of Commercial Polymers.

Dedication

Dedicated to my paternal friend and physicist colleague
Dr. Gustav Richter

Konfuzius *(551–479 BC)*
Humans have three ways to act intelligently:
First of all: by thinking – that is the noblest.
Secondly: by copying – that is the easiest.
Thirdly: by experience – that is the bitterest.

Foreword

In the last decade, technical progress has positively influenced the methods of dynamic mechanical analysis. At the same time, data handling has become more comfortable and much easier.

In this context it is not at all surprising that various techniques of rheology have opened up new insights into so far unknown and undiscovered structures.

Furthermore, new correlations between rheological behavior and specific product or structure properties have been revealed and are used for systematic investigations.

However, sound information about the proper use of rheological techniques is still weak. The majority of published information deals with the rheology of polymers. This book focuses on the rheology of dispersions and emulsions. Students, chemists, engineers and laboratory assistants working on these materials, will find in this book fundamental principles, how to apply rheology, and what kind of information can be obtained.

I wish all readers a lot of energy and enthusiasm for the opportunities offered by rheological techniques.

May 2005

Prof. Dr. Klaus-Peter Wittern
Corporate Vice President
University of Hamburg, Department of Chemistry

Preface

In the last 20 years, personal computers have become more and more powerful. As a result, dynamic mechanical analysis (DMA) has become more and more efficient and rheology has consequently become a common tool in the analytical laboratory. Modern rheometers today are three times cheaper than 10 years ago but four times better in performance. Now this technique and the powerful PC are more often employed by the non-specialist.

However, information on the use of this technique is still thinly scattered. There are several excellent books about rheology and many papers covering correlation with other techniques. Most of these describe polymers and only a few books refer to dispersions and emulsions. Still today you often hear the question: "What is DMA and what can it tell me about my product or process?"

This book attempts to give students, chemists, engineers, and laboratory assistants in the cosmetic field a starting point to understand where and how rheology can be applied. Therefore I have minimized the mathematics and statistics and have given information on how to use a rheometer. Rheology is an efficient tool for getting information on material behavior under different conditions and it can be done very cost effectively when done properly.

Hamburg, May 2005 Rüdiger Brummer

Acknowledgements

I have so many people to thank for their help and support; more than I can list here. First of all, I would like to thank Prof. Kulicke for his suggestion and Prof. Wittern for his encouragement to write this book. They gave me the motivation for this project.

Special appreciation is expressed to my colleagues in the Rheology and Thermo Analysis Laboratory – Frank Hetzel, Martin Griebenow, Rüdiger Uhlmann, Volker Schlesiger and Angelika Wiese – for their collaboration and careful preparation of all the test specimens, since all measurements were done in our laboratory.

I would also like to thank all the students who finished their studies in my laboratory, especially Dr. Thorsten Berg, Dr. Sybille Friedrich and Dipl. Ing. Mandy Mühl, for their dedication and the results of their work, some of which I was able to use in this book.

For the micrographs I would like to thank Dr. Roger Wepf and his coworkers at Beiersdorf. All other figures were taken from the manual of the rheometer supplier, or from internet portals, or are my own.

Finding the best English words was the task of Dr. Marcia Franzen-Hintze, who showed a great propensity to understand my point of view on rheology.

I am also grateful to Prof. Werner-Michael Kulicke and Dr. Christian Clasen, who were kind enough to review this manuscript.

Special thanks go to my friend and fellow rheologist, Dr. Bernhard Hochstein, for stimulating discussions while interpreting the data and for his help in reviewing the formulas.

Last but not least, I would like to thank my family and especially my wife, who was so tolerant and understanding while I was writing, revising and correcting this book on holidays, weekends, evenings, etc.

Hamburg, May 2005 Rüdiger Brummer

List of Symbols

A	Space	m^2
b	Mean droplet diameter	m
c	Concentration	$mol\,l^{-1}$
C_1, C_2	Coefficient	–
d	Diameter	m
$d_{v,10}$	10% of the volume diameter	m
$d_{v,50}$	50% of the volume diameter	m
$d_{v,90}$	90% of the volume diameter	m
E_A	Activation energy	$J\,mol^{-1}$
E/m	Energy input	$J\,kg^{-1}$
F	Force	N
G_E	Modulus of an ideal elastic solid	Pa
G^*	Complex modulus	Pa
G'	Storage modulus	Pa
G''	Loss modulus	Pa
G_p	Plateau modulus	Pa
$G_{1\,rad/s}$	Storage modulus at $\omega = 1\,rad/s$	Pa
h	Thickness	m
I	Current	A
L	Length	m
M	Molecular weight	$g\,mol^{-1}$
M_{cp}	Torque for cone plates	$N\,m^{-1}$
M_{pp}	Torque for parallel plates	$N\,m^{-1}$
n	Revolutions per minute	rpm
p	Pressure	Pa
p_1	Intake pressure	Pa
p_2	Outtake pressure	Pa
Q	Volume per time	$m^3\,s^{-1}$
R	Radius	m
Re	Reynold number	–
r	Radius	m
t	Time	s
T	Temperature	°C
T	Absolute temperature	K

U	Voltage	V
v	Speed	$m\,s^{-1}$
$\dot{V}$	Volume per time	$m^3\,s^{-1}$
w	Characteristic rate	s^{-1}
$\overline{x}$	Average length	m
β	Angle	°
δ	Phase angle	°
η	Dynamic viscosity	Pa s
η_{rel}	Relative viscosity	Pa s
λ	Wavelength	m
ϱ	Density	$kg\,m^3$
γ	Deformation	%
$\dot{\gamma}$	Shear rate	s^{-1}
τ	Shear stress	Pa
τ_{yield}	Yield stress	Pa
τ_i	Relaxation time	s^{-1}
ν	Cinematic viscosity	$m^2\,s^{-1}$
ω	Frequency	$rad\,s^{-1}$

List of Abbreviations

ASTM	American Society for Testing Materials
cmc	critical micelle constant
DAB	Deutsches Arzneibuch
DIN	Deutsche Industrie Norm
INCI	International Cosmetic Ingredients Dictionary
ISO	International Organization for Standardization
JSA	Japanese Standards Association
NMR	Nuclear magnetic resonance
PFGSE	Pulsed-field gradient spin echo
PGPH	Polyglyceryl-2-dipolyhydroxystearate
Re	Reynolds number
TEM	Transmission electron microscopy
TGI	Polyglyceryl-3-diisostearate
UWG	Gesetz gegen den unlauteren Wettbewerb
LBMG	Lebensmittel- und Bedarfsgegenständegesetz
HWG	Heilmittelwerbegesetz
MBO	Musterberufsordnung der Deutschen Ärzte

Table of Contents

1 Introduction

Cosmetic emulsions exist today in many forms for the widest variety of applications, including face and hand creams for normal, dry or oily skin, body milks and lotions, and even sun products. Keeping track of them all is not always easy despite product names or parts of them (e.g. hand or face cream) that clearly indicate their use and properties. The author has undertaken to describe these properties and find ways to measure them. His primary field of interest is the area of flow and flow properties. To the specialist, flow is the continuous deformation of a material when a force is applied. The response of a material to a certain deformation and the mathematical and scientific description, explanation and measurement of this response comprise the field of rheology. This book focuses on the application of rheological measurements to cosmetic emulsion and the correlation of these results with data from other tests.

When experts speak of emulsions they mean a blend of substances that cannot normally be mixed. Fats and oils do not dissolve in water but since both oil and water are very useful for the care and cleansing of skin, people have wanted to combine them since ancient times, and in nature they discovered with milk how the basically impossible is possible. "Cleopatra's bath" in "donkey milk" [1], which took place in about 50 BC, is a well-known example.

Under a microscope the fat droplets in milk can be seen floating in water (oil-in-water emulsion). This explains why donkey milk was already a popular skincare preparation in antiquity, but it was also more. Cleopatra knew that donkey milk cleansed as well as cared for the skin. The most striking property of milk is its ability to remove water-soluble and fat-soluble impurities from the skin while replenishing it with oil-soluble and water-soluble skincare substances. That is why donkey milk today still meets many requirements placed on a skincare product.

Natural milk consists of not just the two components oil and water. It is a very complex system in which oil droplets can exist in the aqueous whey only with help from substances referred to as stabilizers such as phospholipids and proteins (casein). These stabilizers concentrate at the interface between oil droplets and water and thereby prevent the oil droplets from "merging" (or "creaming").

This returns us to flow. The description of flow and flow properties has also come down from antiquity. The saying "Panta rei" or "everything flows" (it is just a matter of time) is known [2] to come from the philosopher Heraclitus (Fig. 1.1), who is supposed to have lived around 500 BC. More precisely, "rei" comes from the

Fig. 1.1. Greek philosopher

Greek word rheos = flow. Rheology is therefore the science dealing with flow. But what actually is flow? That is a question we still need to answer!

Who or what is a rheologist and how does one become a rheologist?
There is no professional training or course of studies leading to certification as a rheologist, as rheology deals with many areas of **mechanics**.

Mechanics is defined as the science of the action of forces on the mass particles of matter.

Rheology can be further sub-divided into the following areas:

Kinematics: the laws of motion
Dynamics: the relation of force and motion
Statics: deals with objects in equilibrium

A rheologist therefore studies, among other things, continuum mechanics, including:

- The **mathematical description** of states of tension and deformation (tensors);
- **Phenomenology**, the qualitative and quantitative description of rheological processes;
- **Viscosity**, the interplay of elastic and viscous mechanisms in matter as a function of temperature and time;
- Structural rheology, the substance-specific interpretation of rheological relationships, and
- Rheometry, the actual static and dynamic measurements.

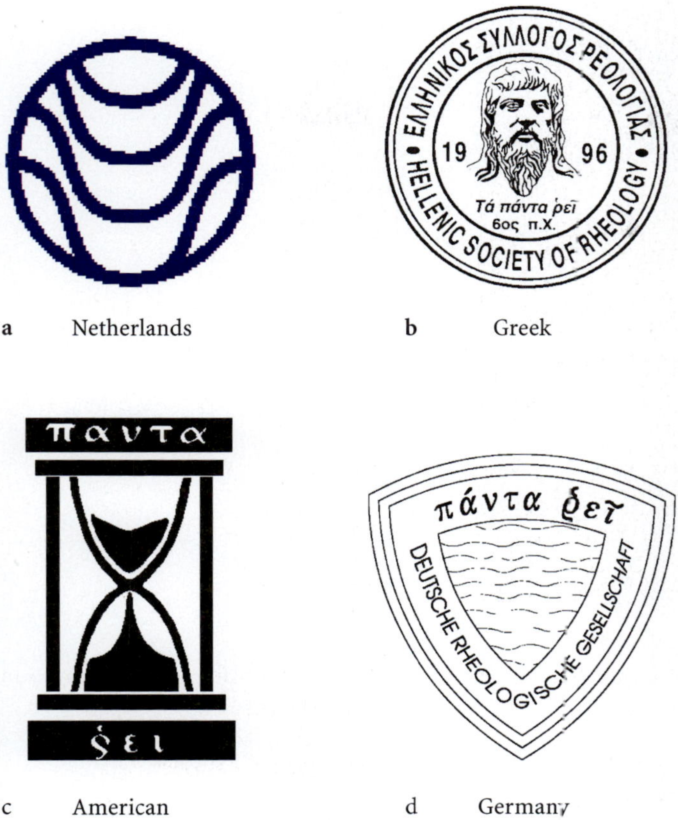

Fig. 1.2. Logos of several rheological societies

A rheologist has studied all these things and it helps for him to have a basic knowledge of chemistry (from a physicist's standpoint) as well. Consequently, it is not unusual to find mathematicians, engineers, physicists and chemists among the "small family" of members in rheological societies (Fig. 1.2) throughout the world.

2 A Trip Back in Time

Let's travel together back in time. About 15 million years ago our universe came into being. The big bang theory now accepted as the standard model [3] assumes that the universe was originally concentrated in an infinitely small space at an infinitely high temperature and density (cosmological singularity) and then made to explode suddenly. This resulted in a rapid drop in temperature followed by the radiation era, which lasted almost one million years. During this period mostly hydrogen, deuterium and helium were formed. Matter was not fully ionized and was still coupled to the predominating radiation field. According to the expansion theory, it was not until much later that the dense matter field decoupled from the radiation field. The pull of gravity then caused the galaxies and stars to form.

Our small planet with its seething magma interior was also formed at this time. From experimental results it is known that this magma forms granite-like melts at temperatures around 700 °C and extremely high pressures. The magma flow depends on its chemical composition and temperature and increases with increasing SiO_2 or Al_2O_3 content. During volcanic eruptions the magma surfacing as lava (Fig. 2.1) has temperatures of 1200 °C and higher. High pressures push it to the surface where it flows down the side of the crater, cooling continuously until the magma ultimately solidifies.

In the course of its evolution the earth has experienced several ice ages. Much is known about the glaciations [4] of the last ice age, which occurred about 20 000 BC. These glaciations left a lasting imprint on the earth's surface. Consisting of millions upon millions of small and large ice crystals, a glacier (Fig. 2.2) moves relatively

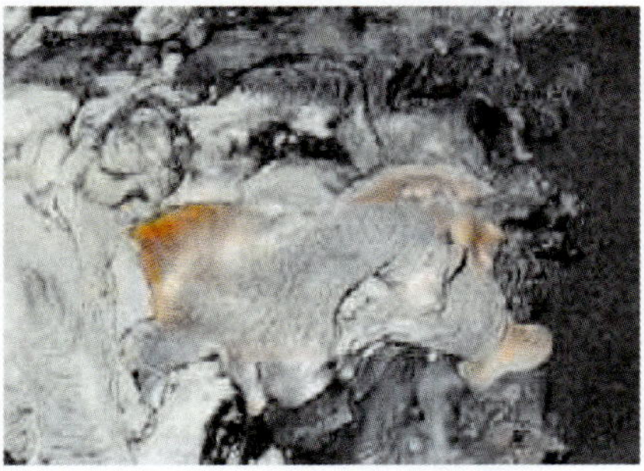

Fig. 2.1. Lava stream

slowly when the temperature is constant. It starts to flow only when the temperature increases and it begins to melt. Ice becomes water rushing to the sea, for instance as waterfalls (Fig. 2.3).

The ratio of water to land surface area is approximately 70:30 or more precisely, the earth as we know it has an area of approximately 510 million km^2, of which approximately 360 million km^2 is water and approximately 150 million km^2 is land.

In each of the examples taken from the earth's evolution, the interplay of temperature, pressure and time produced a change in a property of matter; molten magna (lava) turned to stone, glaciers carved out the landscape and ice crystals became water. It is precisely these three parameters: temperature, pressure and time, as well as the rate of change resulting from their interplay, that interest rheologists.

Fig. 2.2. Glacier

Fig. 2.3. Waterfall

Fig. 2.4. Leonardo da Vinci

It was not until ca. 500 years ago, very recently when compared to the age of the earth, that one man in particular [5], Leonardo da Vinci, studied nature and the laws of nature (Fig. 2.4). To many he is known as an artist, but in addition to being a painter, architect and sculptor, he was also a master of many other disciplines, all of which, however, related to his painting.

The philosopher Leonardo da Vinci was convinced that the sense of sight provides the most unerring picture of reality and therefore painting is the absolute art and science. In his paintings he tried to depict the invisible forces of nature like wind and currents. He first formulated his theories on paper as sketches. Comparison with reality then followed in practical experiments. As an engineer he placed great value on translating his findings into practice.

As the engineer responsible for the waterways of the Po plane, Leonardo da Vinci worked with the element water, which he felt had life-giving as well as life-destroying properties. He studied river currents and how they changed around obstacles such as bridge piers and the resulting eddies. However, his main interest was the erosion of river beds and its prevention.

In 1624, the French mathematician and physicist [6] Blaise Pascal (Fig. 2.5) discovered the law of communicating tubes and demonstrated the decrease in atmospheric pressure with the altitude by measurements with a barometer. To honor his achievements the unit of pressure was given his name.

$$\text{pressure} = \text{force per area}$$

$$1\,\text{Pa} = 1\,\text{N/m}^2$$

In the same century Sir Isaac Newton (Fig. 2.6), the founder of classical theoretical physics [7], formulated the fundamental law of rheology, which was named after

Fig. 2.5. Blaise Pascal

Fig. 2.6. Sir Isaac Newton

him as the Newtonian Law:

$$\tau = \eta \cdot \frac{v}{t} \tag{2.1}$$

This states that the force per area applied to a liquid is proportional to the resulting rate of flow (later called shear rate). The proportionality constant is called the viscosity.

The German engineer Hagen and the French physician Poisseuille [8] studied independently the flow of liquids through tubes in 1839 and 1840, respectively. One approached the problem from a technical viewpoint, and the other wanted to learn how blood flows through arteries and veins.

The law, expressed by the formula,

$$Q = \frac{\pi \cdot r^4 \cdot \Delta p}{8 \cdot \eta \cdot L} \tag{2.2}$$

was named the Hagen–Poisseuille's Law in their honor.

The French physicist C.L.M.H. Navier [9] and the British mathematician and physicist Sir G.G. Stokes published the fundamental differential equation named after them:

$$\varrho \frac{\mathrm{d}v}{\mathrm{d}t} = f \cdot \operatorname{grad} p + \eta \Delta v + \frac{1}{2} \eta \operatorname{grad} \operatorname{div} v \,. \tag{2.3}$$

It describes the general movement of Newtonian fluids in the special case that the dynamic viscosity is constant.

In 1883, the British physicist Osborne Reynolds formulated the law [10] of hydrodynamic similarity in the presence of pressure, frictional and inertial forces. In 1886 he formulated the theory of lubricant friction and effect, which was later expanded by A. Sommerfeld, and in 1889 the theory of turbulent flow followed [11]. The Reynolds number is a dimensionless constant that characterizes the ratio of inertial to viscosity forces in a flowing fluid and is expressed as follows:

$$\mathrm{Re} = \frac{w \cdot L}{\upsilon} \,, \tag{2.4}$$

where w is a characteristic rate, L is a characteristic length (e.g. tube diameter) and ν is the cinematic viscosity.

The next crucial step was the establishment of the word rheology to represent the science of deformation and flow. On December 9, 1929 the American Rheology Society was officially founded. Published just 7 years later in 1936, what is presumably the oldest DIN standard called "Testing of Lubricants" was published (Fig. 2.7).

76 Verbindlich ist nur die neueste Ausgabe dieser Norm im Dinformat A 4.

Prüfung von Schmiermitteln Zähigkeit (Viskosität)	DIN DVM 3655

Begriff

1 Zähigkeit (Viskosität) oder innere Reibung ist die Eigenschaft einer Flüssigkeit, der gegenseitigen Verschiebung zweier benachbarter Schichten einen Widerstand entgegenzusetzen.

Fig. 2.7. The first DIN relating to viscosity

It was not only in Germany [12] that such standards were provided. Internationally more important are the American, Australian and Japanese standards. The need for these ASTM standards [13] and the national standards arose at the beginning of the last century with the development of an increasing number of instruments to measure viscosity. To be able to compare results certain boundary conditions had to be defined and maintained.

ASTM International is one of the largest voluntary standards development organizations in the world – a trusted source for technical standards for materials, products, systems, and services. Known for their high technical quality and market relevancy, ASTM International standards have an important role in the information infrastructure that guides design, manufacturing and trade in the global economy.

ASTM International, originally known as the American Society for Testing and Materials (ASTM), was formed over a century ago, when a forward-thinking group of engineers and scientists got together to address frequent rail breaks in the burgeoning railroad industry. Their work led to standardization on the steel used in rail construction, ultimately improving railroad safety for the public. As the century progressed and new industrial, governmental and environmental developments created new standardization requirements, ASTM answered the call with consensus standards that have made products and services safer, better and more cost-effective. The proud tradition and forward vision that started in 1898 is still the hallmark of ASTM International.

Today, ASTM continues to play a leadership role in addressing the standardization needs of the global marketplace. Known for its best in class practices for standards development and delivery, ASTM is at the forefront in the use of innovative technology to help its members do standards development work, while also increasing the accessibility of ASTM International standards to the world.

ASTM continues to be the standards forum of choice of a diverse range of industries that come together under the ASTM umbrella to solve standardization challenges. In recent years, stakeholders involved in issues ranging from safety in recreational aviation, to fiber optic cable installations in underground utilities, to homeland security, have come together under ASTM to set consensus standards for their industries.

Standards developed at ASTM are the work of over 30 000 ASTM members. These technical experts represent producers, users, consumers, government and academia from over 100 countries. Participation in ASTM International is open to all with a material interest, anywhere in the world.

It was not only in America and Europe that such organizations were founded. In Asia, the Japanese Standards Association (JSA) was set up [14]. The JSA is an organization formed through the merger of the Dai Nihon Aerial Technology Association and the Japan Management Association was authorized to incorporate by the Minister of Trade and Industry on December 6, 1945. Its office was first established at the Patent and Standards Bureau in Chiyodaku, Tokyo, and then moved to Akasaka, Minatoku in 1962. The objective of the association is "to educate the public regarding the standardization and unification of industrial standards, and thereby to contribute to the improvement of technology and the enhancement of production efficiency".

JSA actively participates in ISO and IEC work to develop international standards and directly and indirectly supports the activities of these international standardizing bodies. JSA sends representatives to serve on high level committees at these organizations and provides financial assistance including travel and participation fees for attending meetings, as well as financial and other support to other organizations involved in deliberating draft international standards.

Known worldwide is the International Organization for Standardization (ISO) [15]. International standardization began in the electrotechnical field: the International Electrotechnical Commission (IEC) was established in 1906. Pioneering work in other fields was carried out by the International Federation of the National Standardizing Associations (ISA), which was set up in 1926. The emphasis within ISA was laid heavily on mechanical engineering. ISA's activities came to an end in 1942. In 1946, delegates from 25 countries met in London and decided to create a new international organization, the aim of which would be "to facilitate the international coordination and unification of industrial standards". The new organization, ISO, officially began operations on February 23, 1947.

Because the International Organization for Standardization would have different abbreviations in different languages (IOS in English, OIN in French for Organisation internationale de normalisation), it was decided at the outset to use a word derived from the Greek isos, meaning equal. Therefore, whatever the country, whatever the language, the short form of the organization's name is always ISO.

ISO is a network of national standards institutes from 146 countries working in partnership with international organizations, governments, industry, business and consumer representatives. A bridge between public and private sectors.

Why do we need standards? If there were no standards, we would soon notice. Standards make an enormous contribution to most aspects of our lives – although very often, that contribution is invisible. It is when there is an absence of standards that their importance is brought home. For example, as purchasers or users of products, we soon notice when they turn out to be of poor quality, do not fit, are incompatible with equipment we already have, are unreliable or dangerous. When products meet our expectations, we tend to take this for granted. We are usually unaware of the role played by standards in raising levels of quality, safety, reliability, efficiency and interchangeability – as well as in providing such benefits at an economical cost.

As a follow up of the development of new measuring instruments more and more standards were provided.

Ford cup, Falling ball, Visco balance, Ubbelohde, Capillary viscometer

The simplest instrument is the famous Ford flow cup (Fig. 2.8). The time it takes for a defined volume of fluid to flow through a tube of defined dimensions is measured. Since the temperature of the cup cannot be controlled, the constancy of the temperature is problematic with this method. Temperature control was first introduced with the falling ball viscometer and later instruments.

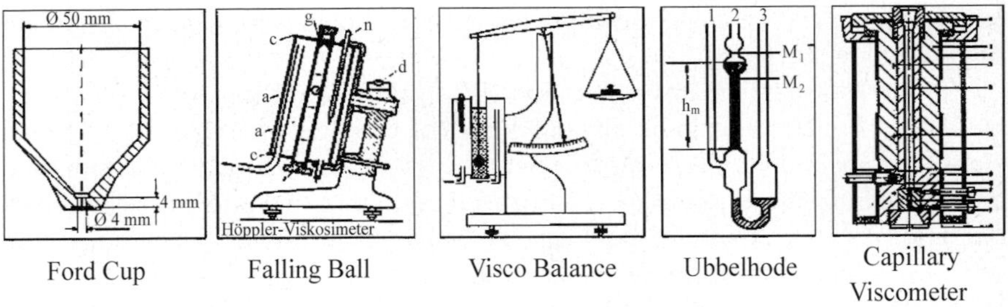

Fig. 2.8. Drawings of some typical viscometers

Advancements in test instruments included the Engler viscometer (Fig. 2.9), the falling ball viscometer and the Ubbelohde viscometer. Whereas only single measuring points can be measured with the instruments first named, entire measuring curves are provided by the latter instruments.

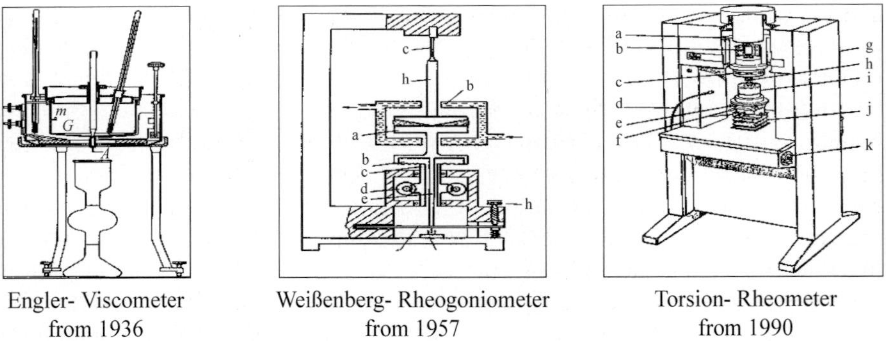

Fig. 2.9. Drawings of selected measuring instruments

Advances in microelectronics in recent decades have allowed the design of instruments that make it possible to record measured values at different rotational speeds. These instruments include the Rheomat or Brookfield instruments and capillary viscometers.

In 1990, the first rotational rheometer was constructed in which a plane shear is measured by transmission to two plane-parallel plates. This was the beginning of the era of the rheometer that not only allow temperature control but also variation of the rotational speed. Progress continued with the creation of the oscillation rheometer that measure the tiniest deformations and speeds and allows determination of the viscoelastic properties of samples. Modern rheometers are able to measure forces ranging from less than 1 mNm to more than 1 Nm in a temperature range from −150 °C to 300 °C.

In retrospect, it can be said that modern rheology has its origins in the seventeenth century. The theories formulated then are still valid today. They have been

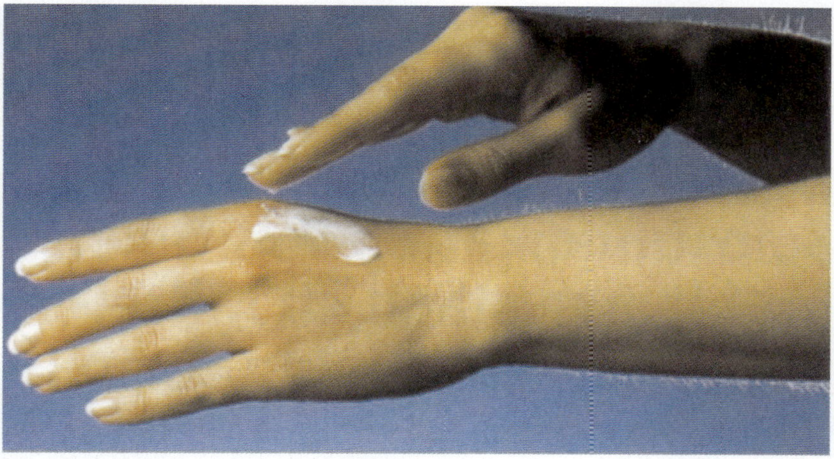

Fig. 2.10. The universal rheometer: the hands while applying a cream

and are still being extended by scientists the world over for application to special problems. However the breakthrough of rheology was only possible with rapid growth of the field of electrical engineering.

Progress continues. The first microrheometer has been constructed and it is only a matter of time until nanotechnology becomes an integral part of the field of rheology. At the beginning of the twenty-first century the focus is on combining instruments, for example a rheometer with DSC cells or with dielectric spectroscopy, but other possibilities are optical systems that allow video recordings of emulsion droplets during shearing.

Engineers have repeatedly attempted to construct a universal rheometer that takes temperature into account and can simulate and measure minute and large deformations as well as weak and strong forces. But why not invent an instrument that already exists and every person has, with the emphasis on *every*? You are now probably wondering just what this instrument is. The answer is quite simple. It is a tool used every day to lather up and then cream the skin or to spread shaving cream. The universal rheometer is our hand (Fig. 2.10)! Even if the perfect measuring instrument were constructed to test an emulsion, it is the consumer who ultimately decides whether a product spreads well, has the right consistency and feels pleasant on the skin. And how do consumers decide this? By spreading an emulsion on their skin with their hand.

3 Skin and its Care

Human skin needs care From the moment of our birth our skin begins to age, as does the whole body, in a natural physiological process [16]. Biological skin aging begins from about the age of 25. Physiological skin aging is accelerated by many external factors like sunlight, cold, UV radiation and air pollution.

Too frequent cleansing depletes the skin of its intrinsic components like horny cells, skin lipids and water. In addition, today's diet, lack of exercise, too little sleep, stress and improper care also affect premature skin aging.

It is with good reason we say the face is the mirror of our soul. We can tell from a person's face whether he is healthy or happy. Therefore everybody will take care of the facial skin by cleaning, resurfacing and moisturizing. The skin is one of our most complex organs [17]. To stay beautiful and healthy it needs modern treatment methods as well as care and relaxation.

With a total area of approximately 2 m^2, the skin is the largest organ of the human body, and it has roughly 4 million receptors (antennas). These are nerves that help us perceive cold and heat and feel pain.

Unlike most cells of the body, which no longer divide once they have matured, skin cells continue to divide throughout their entire lifetime. The skin renewal process takes about 28 days. Continuously forming new cells, the cells in the above layers are pushed increasingly upwards to the surface (Fig. 3.1), where they slowly dry out and form the uppermost horny layer of the skin.

As the external boundary of the body, the skin has several functions. Among its major functions are protection of the body, regulation of body temperature and sensory perception. To ensure these diverse functions can be fulfilled, healthy skin has a natural protective system consisting of secretions from the sebaceous and sweat glands, the skin's own moisturizing factors, as well as amino acids and lactic

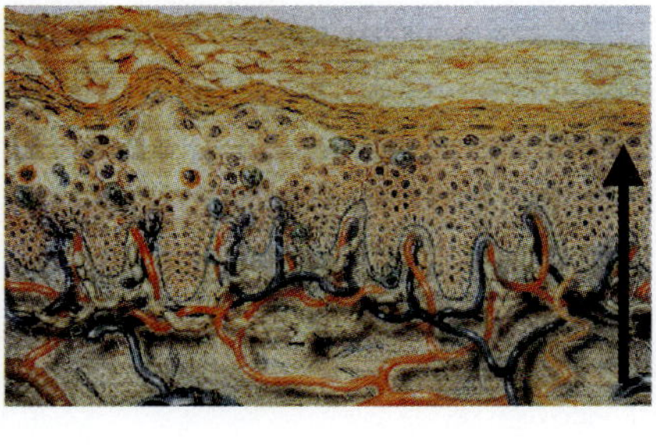

Fig. 3.1. Schematic representation of the skin with the top horny layer

Fig. 3.2. Creaming sun lotion to the skin

acid. This so-called protective acid mantle covers the surface of the skin like an invisible extremely thin film and has a pH that varies between 5 and 6. This is why pH plays an important role in skin cleansing. Products with a pH in this range are said to be neutral or skin friendly.

The protective acid mantle of the skin is influenced by sebum and sweat production. If acids predominate the skin will be dry and feel tight. A predominance of bases will result in oily skin. An important task of skin care is therefore to restore the natural balance of acids and bases. Cosmetic emulsions (Fig. 3.2) play an important role in this arena.

4 Emulsions – Some Theoretical Aspects

The theoretical background for the rheological measurement of emulsions especially for cosmetic emulsions will be presented in this chapter. In a separate chapter I will make an excursion into food rheology and explain other types of emulsions, but now we will start with cosmetic emulsions. After the physicochemical structure of cosmetic emulsions are explained, the rheological principles and rheological test methods needed to measure them will be discussed.

4.1 Physicochemical Structure of Cosmetic Products

The main purpose of cosmetic products is to supply the skin with lipids and moisture. In the field of medicine the purpose can also be to supply active ingredients that must be applied sufficiently diluted in a cream to diseased skin areas. The principle components, however, are always water and oil. Since water and oil are hardly miscible, other ingredients are needed to make them mix. These may be emulsifiers or surfactants that ensure the stability of oil droplets dispersed in water or vice versa or they may be polymer molecules that stabilize emulsions by forming a three-dimensional network in which oil droplets can become interspersed.

The following categories of currently manufactured cosmetic products were defined by Brandau [18]:

- Ointments
- Creams
- Gels
- Lotions

Ointments are spreadable, non-transparent formulations at room temperature that are virtually water-free. They comprise only a minor portion of cosmetic products. Creams differ from ointments in that they consist of fat-like substances, water and usually emulsifiers. Creams can in turn be sub-classified by the emulsion type. In lipophilic creams water is the dispersed and oil the continuous phase. This type of emulsion is abbreviated as W/O. Conversely, hydrophilic creams have oil as the dispersed and water as the continuous phase and are called O/W type emulsions. Amphiphilic creams have both lipophilic and hydrophilic properties. Gels are spreadable, transparent formulations at room temperature, whereas lo-

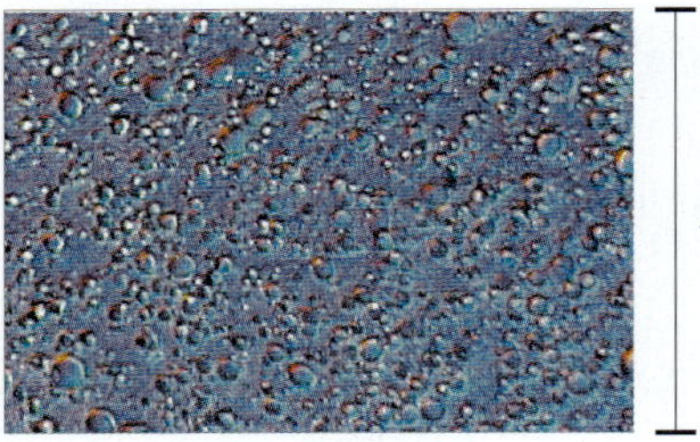

Fig. 4.1. Photomicrograph of an O/W emulsion

tions are free flowing creams (mainly of the O/W type) at room temperature. The droplet diameter of the disperse phase usually ranges from 1 to 5 µm, as shown in Fig. 4.1.

Other possible emulsion types are W/O/W and O/W/O formulations. The oil droplets in a W/O/W emulsion are emulsified in water and the water droplets in turn emulsified in the oil droplets, as can be seen in Fig. 4.2. The size of the oil droplets ranges from 5 to 10 µm and that of the water droplets from approximately 1 to 2 µm. The opposite is true of O/W/O emulsions.

Emulsions are thermodynamically metastable systems exposed to physical, chemical and microbiological influences during manufacture, transport, storage and use that can produce visible changes in the emulsion. Such changes can be caused by temperature, exposure to light, external pressure, etc. These variables affect the solubility product and this can result in crystallization. If interaction of the ingredients with each other or with the packaging material occurs, this can result in instabilities due to chemical reactions. Yeast, bacteria and molds affect the microbiological stability of the product.

Rheological measurements will be presented that can be used to characterize cosmetic products such as creams, lotions and gels. These are plastic materials characterized by non-Newtonian flow behavior. The onset of flow is product-specific and differs significantly for lotions and creams. On the basis of the critical shear stress at the yield point, the emulsion type can be determined for creams as well as lotions. The onset of flow of W/O emulsions is observed at a considerably lower shear stress than with an O/W emulsion. Gels do not have a characteristic yield point but can be distinguished by a critical shear rate. The recovery time after loading below the yield point is not a product-specific characteristic for creams, lotions and gels but crucial for the reproducibility of measuring results

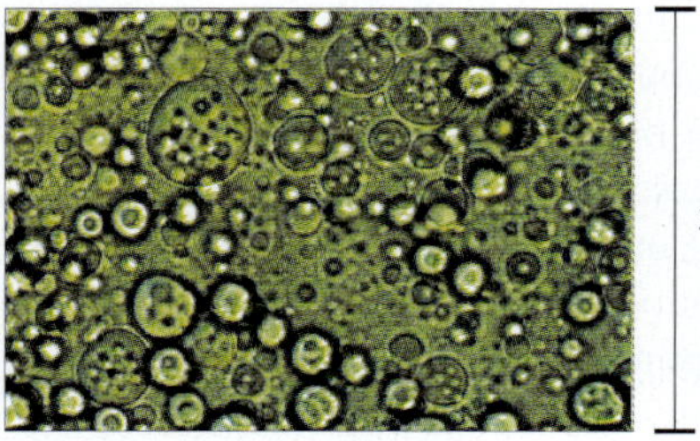

Fig. 4.2. Photomicrograph of a W/O/W emulsion

Cosmetic cleansing products containing surfactants are characterized by Newtonian flow behavior. In this product group no recovery takes place after shearing. When subjected to periodic, usually sinusoidal deformation, hydrogels show typical polymer characteristics. At low frequencies they behave like a fluid and at high frequencies like an elastic solid.

4.2 Modern Emulsifiers

Modern emulsifiers [19] are mainly surfactant additives that reduce surface tension. They include foaming agents, defoamers, wetting agents, detergents and solubilizers. Very different emulsion structures can be achieved depending on the emulsifiers used and their concentration. Consequently, a variety of applications are possible. A bar of ordinary soap consists almost entirely of a pure emulsifier that can absorb fats when combined with water. Consequently, ordinary soap is used to cleanse the skin, i.e. to remove fatty impurities, but also excess sebum. However, the same emulsifier can be mixed with emollient oils, water and water-soluble skincare substances to make oil-in-water (O/W) creams. These skincare creams have long been known as stearate creams and today are occasionally still found in the skin protection sector, for example as products with a high content of free stearic acid. This emulsifier has been replaced mainly by pure synthetic emulsifiers that offer several advantages in terms of their performance characteristics.

4.3 Skin Care and Cleansing

Emulsions look milky-white like natural milk and are incorporated in cleansing creams as well as skincare cream (semisolid) and lotions (liquid). The oil droplets in these O/W emulsions are about 1–20 μm, or 0.001–0.020 mm, in size. Conversely, emulsions may contain water droplets (W/O creams) or even be multiple systems [20] (W/O/W and O/W/O). O/W creams usually supply more moisture and W/O creams more lipids.

The smaller the mean droplet size, the more transparent the products are. Emulsions with a droplet size distribution between 10 and 50 nm are called microemulsions. They are fully transparent and distinguished by a relatively high emulsifier content.

4.4 Microemulsions

Microemulsions [21] are used for different purposes. The high emulsifier content has a strong influence on the skin barrier, resulting in very fast penetration or permeation of active ingredients through the skin. This is especially advantageous in the pharmaceuticals sector for drug therapies. In the skincare sector this proves to be more of a disadvantage because emulsifiers severely disturb the integrity of the skin barrier layers. In cosmetics, microemulsions are used mainly for skin cleansing, e.g. as oil-containing cleansing gels, shower gels and bubble baths.

Microemulsions in the narrower sense are systems with a high surfactant content that actually are not emulsions because the water and oil phases can no longer be discerned even under an electron microscope. They are occasionally used for transdermal drug formulations but due to their emulsifier side effects are no longer of much importance. In contrast, no clear distinction is made between two- and one-phase systems in the area of skin cleansers.

4.5 Emulsifier-Free Products

Whereas emulsions and microemulsions are based on a more classical concept and contain largely synthetic emulsifiers, nanoemulsions are based on a markedly physiological concept. The particles in nanoemulsions are smaller than those in microemulsions, having a diameter from 0.00005 to > 0.0001 mm. Nanoemulsions do not contain typical emulsifiers but rather pure, natural phosphatidylcholine. Phosphatidylcholine, which is obtained from lecithin, is the essential building block of all natural cell membranes. Unfortunately, the INCI name [22] for phosphatidylcholine is lecithin, which makes it impossible for the non-professional to distinguish between the two on the package label. Phosphatidylcholine dispersions spontaneously form bilayer membranes like those of the cell membranes, the barrier layers of the skin and liposomes. Using high-pressure technology it is possible to force phosphatidylcholine to form simple membranes that can enclose oil droplets, making conventional emulsifiers superfluous. Conditions are achieved that resemble those found in the body's own fat transport system, the chylomicrons.

Phosphatidylcholine can be completely metabolized and additionally provides the skin with two essential substances: linoleic acid and choline. Therefore phosphatidylcholine actually has little in common with conventional emulsifiers, and the term nanoemulsion was quickly supplemented or replaced by terms like nanodispersion, nanoparticle or nanoparts. Nanoemulsions are used for example for intravenous fat nutrition. Analogous use of conventional emulsifiers for this purpose would quickly result in destruction of the blood and blood vessels.

In the cosmetics sector, nanoparticles belong to the group of emulsifier-free products. They have one distinct advantage: whereas emulsifiers are usually stored unchanged in the skin and tend to promote washout of the skin's own lipids with the next skin cleansing, phosphatidylcholine shows just the opposite effect. It has an almost magical attraction for lipids into the skin. This is also true of the donkey milk mentioned earlier and observed with balneological products as well. Due to the high production costs, nanoparticles are incorporated in higher amounts only in special products such as products for elderly and problem skin as well as products for supportive preventive care.

Cold cream is possibly the oldest emulsifier-free cream and cream dermal membrane structure (DMS) cream is the newest. DMS cream is not classified as an emulsion as no droplet structures can be seen under the normal light microscope. Lamellar structures like those typical of the barrier layers of the skin become visible only in the electron microscope. DMS cream is made of a phosphatidylcholine that contains esters of the palmitic acid and stearic acid predominating in the horny layer rather than of linoleic acid. Interestingly, they have properties similar to those of ceramides. They anchor themselves in the barrier layers of the skin and like ceramides are very resistant to exogenous substances acting on the skin. Consequently, ceramides, DMS, liposomes and nanoparticles are compatible with each other in nearly any ratio. DMS creams cannot be produced using the common emulsification methods although they do not differ from emulsions in their appearance or use. DMS creams are suitable for extremely sensitive and problem skin because they do not disrupt the skin barrier.

4.6 Production of Emulsions

At first glance making an emulsion seems to be a simple process. When two immiscible liquids are dispersed by stirring vigorously an emulsion is obtained briefly. If the two liquids are water and oil, either a W/O or an O/W emulsion will be formed depending on the amounts of each liquid used. Because the free energy of the emulsion system [23] is higher than that of the two liquids, the phases will again separate with release of energy. To stabilize these systems for longer periods emulsifiers must be incorporated that delay phase separation into the thermodynamically more stable starting liquids until after the emulsion has been used as intended.

The emulsion production process can be divided into three basic steps:

1. Pre-emulsification
2. Fine emulsification
3. Stabilization

In the pre-emulsification step the water and oil phases are combined at an elevated temperature with stirring, forming a raw emulsion (premix) with large droplets. These are deformed in the subsequent fine emulsification step by external shear forces and their size reduced when a critical deformation is exceeded. The newly formed interface is then protected by emulsifiers against coalescence in the stabilization step.

4.7 Processes Occurring During Emulsification

The emulsification process entails the breakup of droplets and wetting of the newly formed interface, which is no longer completely covered by emulsifier molecules immediately after size reduction. Adsorption of more surfactant molecules takes time and depends on the interfacial wetting kinetics of the emulsifier system used. The coverage density influences not only the interfacial tension and hence the energy needed for particle size reduction but also the stability of the droplets generated [24].

Insufficiently stabilized droplets can coalesce upon impact with other droplets if the contact time is long enough. For coalescence to take place, the continuous phase between colliding droplets must be displaced to a critical film thickness (film drainage). Coalescence can be prevented if the repulsive forces between droplets are sufficiently high. These repulsive forces are exerted by the adsorbed emulsifier molecules. Spreading of emulsifier molecules unevenly distributed on the droplet surface (Gibbs–Marangoni effect) [25] slows film drainage and stabilizes the droplets even if the interface is not completely covered [26].

Droplet size reduction as well as coalescence of broken up but not yet completely stabilized droplets determine the emulsification results and the dispersity of the emulsion formed.

4.8 Serrated Disc Disperser

Droplet size reduction requires normal and/or tangential tensions at the interface between the internal and external phase. Droplets are broken up when local deforming forces exceed form-retaining interfacial forces for a long enough period. This requires dissipation of large amounts of energy in the dispersing zone of an emulsifier machine.

The serrated disc disperser consists of a rotor-stator system constructed of coaxially intermeshing discs with slots. The width of the gap between the rotor and stator is in the order of magnitude of millimeters. The emulsion, which is placed in the middle of the disperser, is accelerated by the centrifugal force of the moving rotor and decelerated by the stator. The shear forces arising are generally thought to be responsible for droplet size reduction [27]. Serrated disc dispersers are usually self-propelling due to the way their flow is guided.

After dispersion, the emulsion droplets (Fig. 4.3) pass usually in a laminar flow through pipes where they can collide. If the phase interface is not sufficiently stabilized by adsorbed emulsifier molecules and the contact time is long enough for the continuous phase between the droplets to be displaced, the droplets will coalesce. The resistance to coalescence immediately after the droplets are broken up is called the short-term stability [28]. The short-term stability of emulsions is influenced not only by the adsorption kinetics of the emulsifiers but also by the coalescence probability of the droplets. The latter is determined by the contact time and interparticulate interactions.

When two droplets collide the continuous phase between them is displaced (film drainage), i.e. the film ruptures once a critical thickness is reached.

The critical film thickness for emulsions is in the order of magnitude of 1–100 nm [29]. If the film of the continuous phase ruptures, the droplets will coalesce spontaneously. Chesters et al. could show that the coalescence probability depends on the critical film thickness, the viscosity ratio of the dispersed and continuous phase, the droplet radius and the Weber number [30]. They also showed

Fig. 4.3. A view of production

that the coalescence probability in laminar flow is higher than in turbulent flow which means that the droplets in pipelines downstream from the dispersion zone are the most susceptible to coalescence.

Besides the destabilizing mechanisms associated with incomplete coverage of interfaces, there is also a stabilizing effect referred to as self-healing of the interfacial film. The Gibbs–Marangoni effect produces an increase in emulsifier concentration in the contact zone between two incompletely covered droplets. The pressure in the contact zone increases as the concentration gradient levels off, and the droplets are pushed apart [31].

5 Basic Physical and Mathematical Principles

After this short excursion into the basic principles of emulsions we will now take a closer look at the physics or more precisely the mechanics and mathematics of rheology. These are the basic principles rheologists know and use. In other words, we will be looking at some theory, definitions and a few equations.

5.1 Important Definitions

Let's begin with the most important definitions [32], for they are essential to our understanding of this field.

Table 5.1. Definition of flow behavior for T = const.

Newtonian	The viscosity is independent of the shear rate
Structural viscosity	**Broad term** for all non-Newtonian flow phenomena
Pseudoplastic	The viscosity shows **Newtonian flow properties at low shear rates** but the **viscosity decreases** above a critical shear rate
Plastic	The viscosity **decreases** with **increasing** shear rate
Dilatant	The viscosity **increases** with **increasing** shear rate
Thixotropic	The viscosity **decreases** at constant temperature and **constant shear rate** over time and returns to its **original state** in a finite time when the **shear** is **removed**
False thixotropy	The viscosity **decreases** at constant temperature and **constant shear rate** over time and does **not** return to its original state in a finite time when the **shear** is **removed**
Rheopexy	The viscosity **increases** at constant temperature and **constant shear rate** over time and returns to its **original state** in a finite time when the **shear** is **removed**

A shear rate-time profile is programmed at constant temperature. For every shear rate defined the resulting shear stress is measured and used to calculate the viscosity. A constant viscosity value is obtained for substances with ideal viscous behavior (Newtonian flow properties). For substances with pseudoplastic flow properties the viscosity increases with increasing shear rate. Dilatant fluids show an increasing viscosity with increasing shear rate. Usually viscosity curves are recorded with increasing shear rates. However, it is also possible to start at a high shear rates and gradually approach the low shear rate. If both the upward and downward curves are measured for a sample the load-dependent as well as the time-dependent flow properties can be obtained. In practice, the area between the upward and downward curve is often calculated as a measure of the time-dependent flow behavior.

If a substance shows dilatant flow behavior (Fig. 5.1), it thickens when shear stress is applied. As a result, the shear rate (see Chap. 2) increases more slowly than the shear stress. The shear viscosity is not constant but increases. This is due to interactions between hardly solvated substance particles [33] as well as the immobility of the dispersing medium. A starch solution is an example of a substance with dilatant flow properties.

A substance is pseudoelastic if the increase in shear stress induces a disproportionate increase in shear rate. With increasing velocity gradient the viscosity therefore decreases. However, at low shear rates the shear viscosity of a pseudoplastic substance is ideally constant. In other words, it is independent of the velocity gradient. The subsequent viscosity decrease can be explained by structural changes. Structurally viscous fluids contain irregularly shaped particles, droplets or branched and/or entangled long molecular chains. At rest, the entropy is high, i.e. the particles, droplets and molecules are distributed chaotically in the structurally viscous material. The system strives to maintain this state, but if the shear stress is increased further, the structural components align themselves in the direction of flow. Entangled molecular chains detangle and the spherical coil of the macromolecular chain is deformed into an ellipsoid. Also droplets in emulsions take on an ellipsoidal shape and aggregates [34] decompose into their elements (Fig. 5.2). Understandably, the system will flow more readily in a state where its components can align with the direction of flow. However, for semidilute polymer solutions it could be shown that the effect of detanglement of the polymers on the

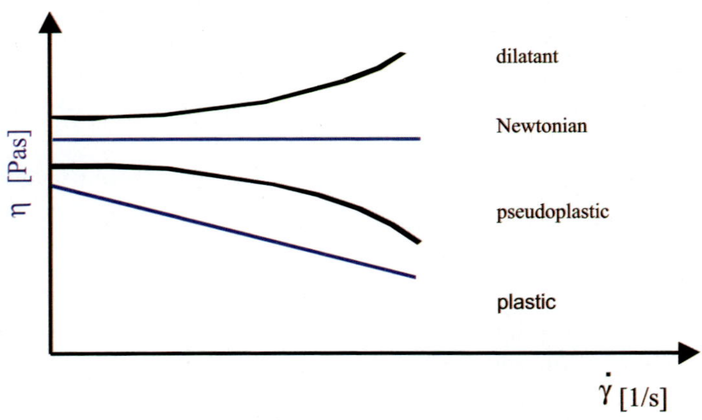

Fig. 5.1. Different viscosity curves

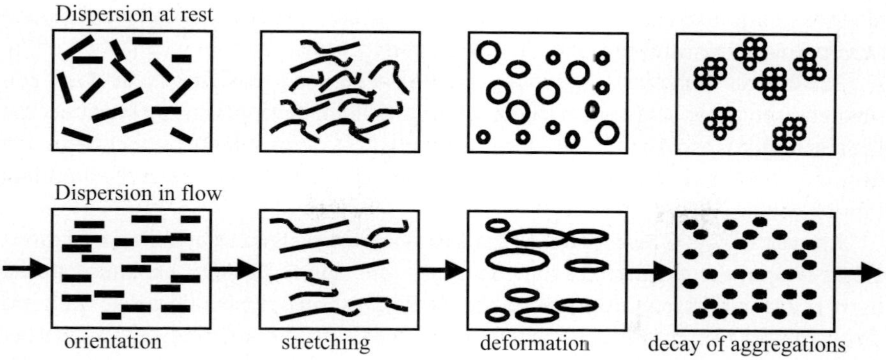

Fig. 5.2. Dispersions at rest and in flow

shear thinning is much more pronounced than the deformation of the polymer coils [35]. Droplets take on an ellipsoid shape and aggregates decompose into their elements (Fig. 5.2). Understandably, the system will flow more readily in this state that is aligned with the direction of flow.

When products are filled they are usually pumped through pipes and are subjected to shear stress. The consumer stresses a product when he presses it out of a tube or spreads it on the skin with his hand by rubbing. Shear rates of at least $10\,s^{-1}$ are attained in both these processes. If we want to study products that are as

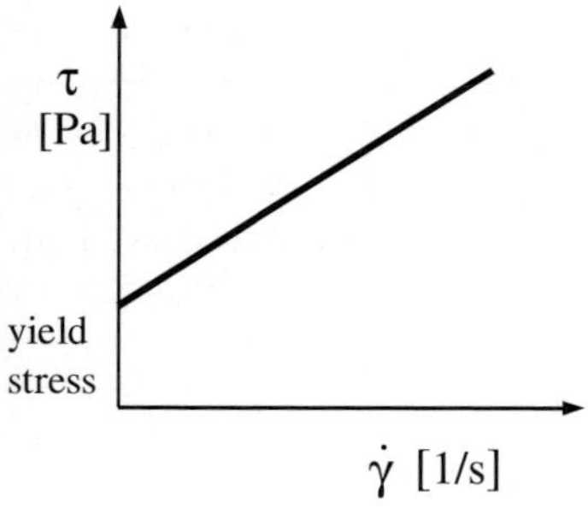

Fig. 5.3. Curve for a Bingham model

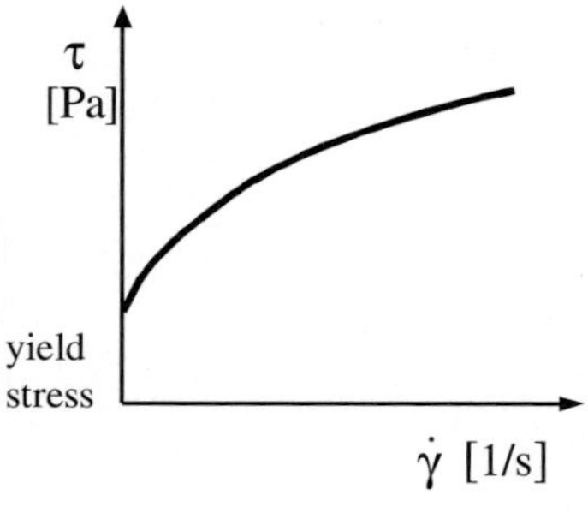

Fig. 5.4. Curve for a Casson material

close to possible to the resting state only very small shear stresses may be applied or very low shear rates of less than $10\,s^{-1}$ defined.

Substances with plastic flow properties have a yield stress. The shear stress can be increased up to a specific value without any deformation taking place because the resistance is too high. If the maximum value is exceeded the substance begins to flow. After the maximum shear stress is exceeded a sharp decrease in the viscosity takes place.

Above the flow threshold a Bingham model is characterized by a linear relationship between the shear stress and the shear rate (Fig. 5.3). For a Casson material there is also a relationship between the shear stress and shear rate above the yield stress, but in this case it is non-linear (Fig. 5.4). Other mathematical models besides those of Bingham and Casson [36] that describe specific sub-regions of the measuring curve include those of Newton, Steiger/Ory, Ostwald.

Dispersions with a high proportion of dispersed phase like emulsions usually exhibit plastic behavior because of various interactions between the dispersed particles. Often a solvation sheath forms around the particles, immobilizing the external phase.

5.2 One-Dimensional Parallel Plates Model

With these definitions in mind we will now turn our attention to the theory. We will start by considering an everyday activity, spreading butter on a slice of bread. We have three starting materials: the slice of bread, the butter and the knife. The bread and knife can be thought of as two flat plates and the butter as a viscous fluid between them. The first step is to spread the butter, which requires a force. The force needed to spread the butter on the bread will depend on how much earlier the butter was taken out of the refrigerator. But what happens to the butter when force is applied? Several things happen simultaneously. The butter on top at the knife moves with the same speed as the knife and is simply left behind on the bread. This is illustrated in simplified form below:

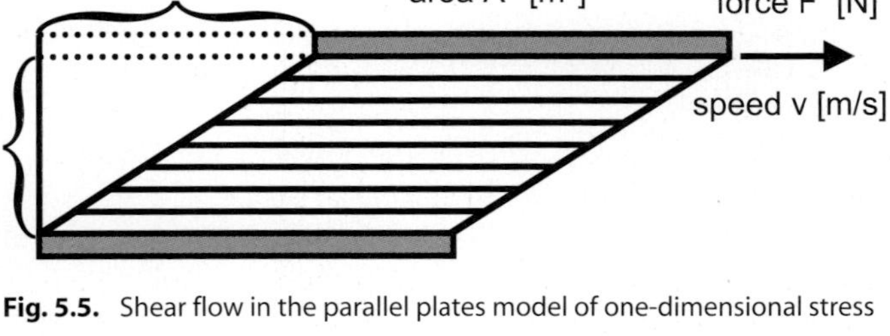

Fig. 5.5. Shear flow in the parallel plates model of one-dimensional stress

We have two [37] plane parallel plates (Fig. 5.5). Located between them in our example is butter of thickness h. The top plate with an area A [m^2] is moved with a velocity v [m/s] by the force F [$N = kgm/s^2$]. Between the two plates a shift in the minute laminar fluid layers takes place. The flow arising is laminar and not turbulent.

The ratio of the force F to the area A is called the shear stress:

$$\tau = F/A\,[\mathrm{N/m^2} = \mathrm{Pa}]\ . \tag{5.1}$$

The ratio of the velocity v to the thickness h is the shear rate:

$$\dot{\gamma} = v/h\,[1/\mathrm{s}]\ . \tag{5.2}$$

The deformation arising is:

$$\gamma = x/h\,[\text{dimensionless}]\ . \tag{5.3}$$

This experiment gives us the following additional information: the shear stress increase is proportional to the shear rate. The proportionality factor was called viscosity by Sir Isaac Newton:

$$\tau = \eta \cdot \dot{\gamma}\,[\mathrm{Pa} \cdot \mathrm{s}]\ . \tag{5.4}$$

This law applies only to a very small category of substances called Newtonian fluids.

The velocity between the parallel plates and Newtonian flow behavior is linear. Depending on the geometry of the measuring system and the sample to be measured the shear rate distribution might not be constant. Then we speak of a non-constant velocity gradient or nonlinear behavior.

Now let us return to our slice of bread. The following happens when butter is spread. First a thick layer of butter is applied to the slice of bread which is then spread evenly over the whole slice. The thickness of the butter decreases with the spreading time. This means that the ratio of the velocity to the thickness is no longer constant assuming the velocity does not change while the butter is spread. However, if the thickness decreases, the denominator will become smaller and consequently the whole amount larger. Therefore the shear rate increases while the butter is being spread on the bread. It is impossible to state a specific value for the shear rate for many processes in our daily lives. Instead a range must be given. Below are several examples [38].

Let us look at another example, rubbing a cream or lotion on the skin (Fig. 5.6). Here, too, the shear rate increases with the cream application time.

Although the time-dependent processes occurring when butter is spread on a slice of bread or a cream emulsion is applied to the skin are very similar, there is one small difference. Whereas spreading butter is a process taking place primarily in one direction, the hand applying a cream uses more or less a closed circular motion. The mathematical model for this is called the torsion gap or subsequently the plate/plate measuring system.

Table 5.2. Examples of typical shear rates

Typical shear rates			
Sedimentation	0.000001	to	0.0001 1/s
Drops from a water faucet	0.0001	to	0.1 1/s
Extrusion	1	to	100 1/s
Spreading butter on bread	10	to	50 1/s
Mixing, stirring	10	to	1000 1/s
Creaming	500	to	10 000 1/s
Pumping	1000	to	50 000 1/s
Spraying, squirting, silk-screening	10 000	to	100 000 1/s

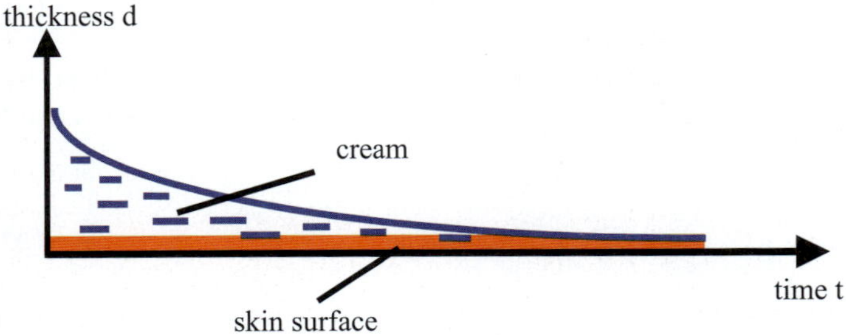

Fig. 5.6. Shear flow during cream application

5.3 Parallel Plate Measuring System

The PP measuring system (Fig. 5.7) has a constant, defined radius R and a variable plate gap h. In DIN 53018 part 1 a plate gap ranging from 0.3 to 3 mm is recommended. The radius R should be several times larger than the gap h.

The angular speed in the gap $\omega(h)$ is constant in levels parallel to the plates and increases with the height:

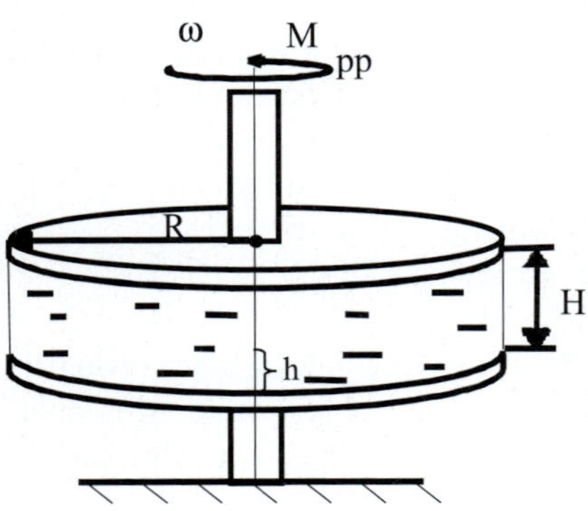

Fig. 5.7. Parallel plate model

$$\omega(h) = \Omega \frac{h}{H} \,. \tag{5.5}$$

The peripheral speed also depends on the height and also on the radius:

$$\omega\left(h\right) = r \cdot \Omega \frac{h}{H} \,. \tag{5.6}$$

From this the shear rate is calculated:

$$\dot{\gamma} = \frac{r\Omega}{H} \,. \tag{5.7}$$

For Newtonian fluids we can calculate the shear stress as:

$$\tau = \frac{2M_{\mathrm{PP}}}{\pi R^3} \,. \tag{5.8}$$

The angular velocity ($\omega = 2\pi \cdot n/60$) is expressed in rad/s and the rate of rotation in min^{-1}. By varying the gap between the plates it is possible to regulate the shear rate. Increasing the gap h (the denominator becomes larger) decreases the shear rate if the angular velocity or rate of rotation remains constant. Care must be taken to ensure that the gap does not become too small because then frictional effects would falsify the measuring results. As a rule of thumb, the gap should be at least five times larger than the largest particles contained in the sample. Consequently, the PP model is most suitable for semi-solid materials and has the added advantage of being easy to clean.

Unfortunately, this measuring system also has one disadvantage. As can be seen from the equation, the shear rate in the PP model depends on the radius. This is not surprising, as the peripheral velocity is zero in the rotation axis and maximum at the rim of the plate at the distance R. This in turn means that in the PP model there is a shear rate based on the maximum radius and therefore the value registered is too large. How this apparent shear rate can be corrected will be explained in Sect. 5.8.1.

5.4 Cone-Plate Measuring System

Since the PP model is more suitable for semi-solid substances and has the disadvantage of variable shear rates, it is legitimate to ask whether there are any measuring systems that do not have this disadvantage and can also measure liquids like water.

If we replace the top plate of the PP model with a cone with its tip pointing towards the bottom plate the result is the cone-plate model (Fig. 5.8), which eventually came to be known as the CP measuring system [39]. This rheometer type is well known in ASTM D4287 for paint and colors and ASTM D3205 for asphalt.

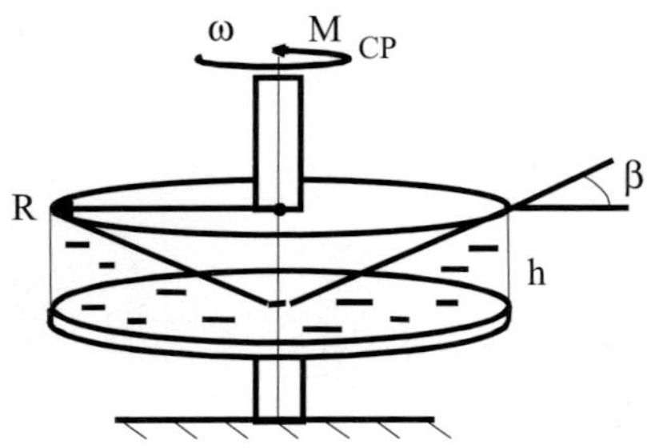

Fig. 5.8. Cone-plate model

This substitution has a surprising effect that is explained below:

Due to the cone angle β the ratio of the corresponding radius to the plate gap is constant for every point on the surface of the cone

$$\tan\beta = \frac{h}{R} \tag{5.9}$$

and for small angles β and consequently $tan\beta$ can always be set equal to β (in radians). This means that the shear rate is constant in the CP model across the entire sample:

$$\dot{\gamma} = \frac{\omega}{\tan\beta} = \frac{\omega}{\beta} \; . \tag{5.10}$$

The shear stress is obtained as in the PP model:

$$\tau = \frac{3 \cdot M_{cp}}{2 \cdot \pi \cdot R^3} \; . \tag{5.11}$$

The cone angle must be small to allow the simplification $\tan\beta = \beta$. To prevent wearing of the cone tip and friction arising from contact with the bottom plate, the cone tip is flattened by 30 to 100 µm. When filling the CP system care must be taken that the distance between the virtual cone tip and the bottom plate is maintained exactly. Usually cones with an angle ranging from 0.5 to 4° are used for measurements. The preferred angle is 1°. Here again the rule of thumb is that the particle size must be five times smaller than the gap (i.e. 6 to 20 µm relative to the virtual gap).

This not only has the advantage that a constant shear rate prevails throughout the gap of the CP measuring system but also allows measurement of relatively high shear rates, small sample amounts and easy cleaning. But the CP model likewise has one minor disadvantage. Liquids like water are very difficult to handle on the bottom plate because they tend to run off the plate (no raised rim). During the measurement, at the latest, the sample will be expelled from the measuring gap by centripetal forces.

5.5 Coaxial Cylinder Systems

Consequently, yet another measuring system is needed to be able to measure any more or less free flowing sample. Once again we will take a very practical approach

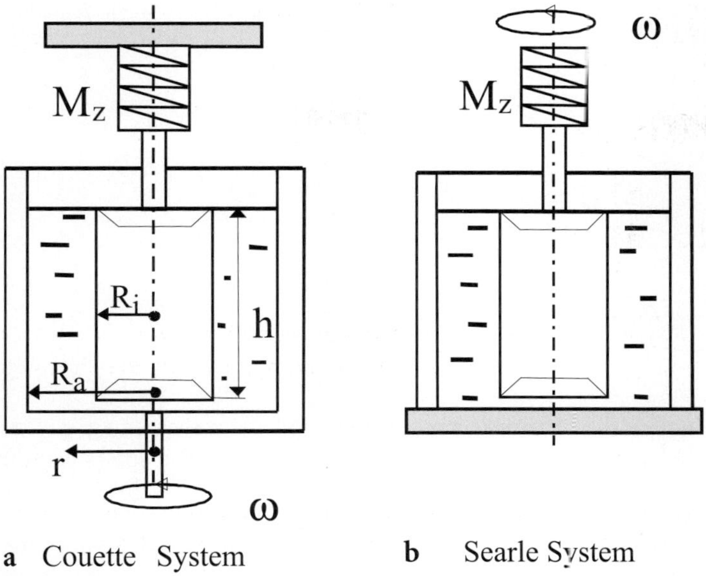

Fig. 5.9. Cylinder systems

and look around for everyday examples. How does a housewife or houseman handle liquids in the kitchen? They are stored in a jar or cup or stirred with a beater or mixing rod. If we apply this image to a rod rotating in a cup, the result is [40] the coaxial measuring system, also known as the concentric cylinder system (Fig. 5.9).

There are basically two types of cylinder systems. One is the Couette system shown in Fig. 5.9a) in which the outer cup is moved and the resulting force measured. The other is the Searle system shown in Fig. 5.9b) in which the outer cup remains fixed and only the inner cylinder rotates and also measures the resulting force.

The definitions from the parallel plate model can be applied to concentric, round, axially symmetrical cylinders (Fig. 5.10) if the surfaces are considered to be infinitesimally small areas.

The freely moving circular area is

$$A = 2\pi \cdot \left(R_{\mathrm{a}}^2 - R_{\mathrm{i}}^2\right) \cdot h \qquad (5.12)$$

This gives the following equation for the shear stress τ:

$$\tau = M_z / 2\pi \left(R_{\mathrm{a}}^2 - R_{\mathrm{i}}^2\right) \cdot h \qquad (5.13)$$

and subsequently the shear rate:

$$\dot{\gamma}_{\mathrm{app}} = \frac{1}{r^2} \cdot \frac{R_{\mathrm{i}}^2 \cdot R_{\mathrm{a}}^2}{R_{\mathrm{a}}^2 - R_i^2} \cdot \omega \qquad (5.14)$$

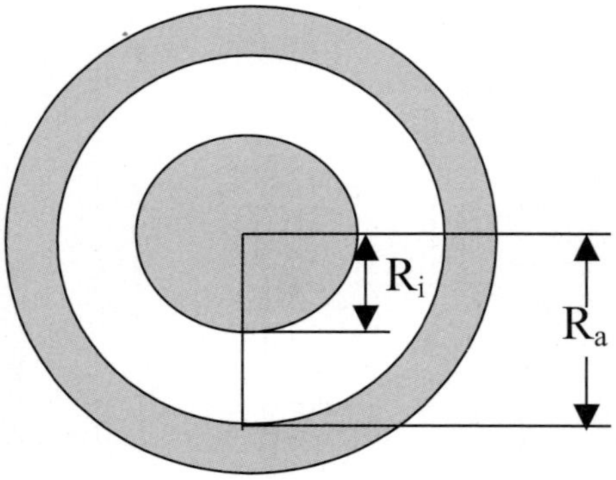

Fig. 5.10. Cross-section of a cylinder system

As in the PP model, the shear rate is not constant in the measuring gap. This is why a correction is again needed, which is indicated by the index app (short for apparent) affixed to the uncorrected shear rate.

DIN 53019/ ISO 3219 defines a maximum radius ratio:

$$d = R_a / R_i \leq 1.1 \quad \text{(preferably 1.0847)} . \tag{5.15}$$

However, the DIN standard does not specify the absolute radii or gap. We will now take a closer look at the schematic drawing in Fig. 5.11:

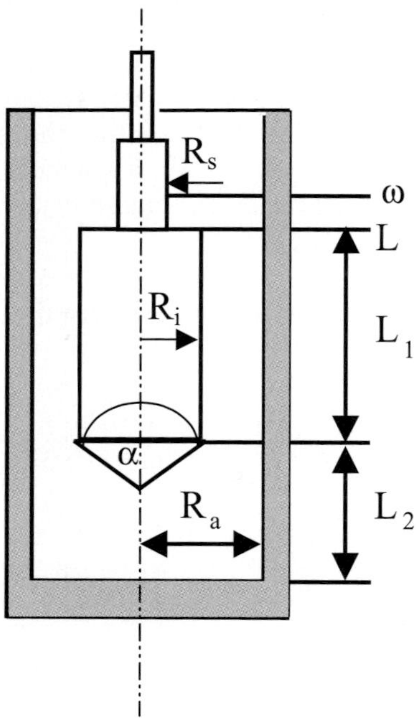

L = Length of the inner cylinder
L_1 = Length of the immersed shaft
L_2 = Distance between the bottom edge of the measuring cylinder and the bottom of the measuring cup
R_i = External radius of the inner cylinder
R_a = Internal radius of the outer cylinder
R_s = Radius of the measuring shaft
α = Angle of the measuring cone

Fig. 5.11. Cylinder measuring system

ISO 3219 specifies the following geometric ratios:

Radius ratios: $d = R_a/R_i \leq 1.1$ (preferably 1.0847)
$R_s/R_i \leq 0.3$
Length to radiuses: $L/R_i \geq 3$ (preferably 3.00)
$L_1/R_i \geq 1$ (preferably 1.00)
$L_2/R_i \geq 1$ (preferably 1.00)
Measuring cone angle: $90° \leq a \leq 150°$ (preferably $120 \pm 1°$)

Therefore the shear stress τ derived from the torque M is

$$\tau = 0.1446 \cdot M/R_i^3 \tag{5.16}$$

and the shear rate obtained with $n = 1/\text{min}$

$$\dot{\gamma} = 1.291 \cdot n \tag{5.17}$$

In extreme cases such as liquids having the consistency of water a double gap system can be used.

5.6 Double Gap Measuring System

This special coaxial cylinder measuring system [41] with a very large shearing area has been standardized for very low viscosities. The actual sample holder is an axially symmetrical gap into which another cylinder is immersed (Fig. 5.12).

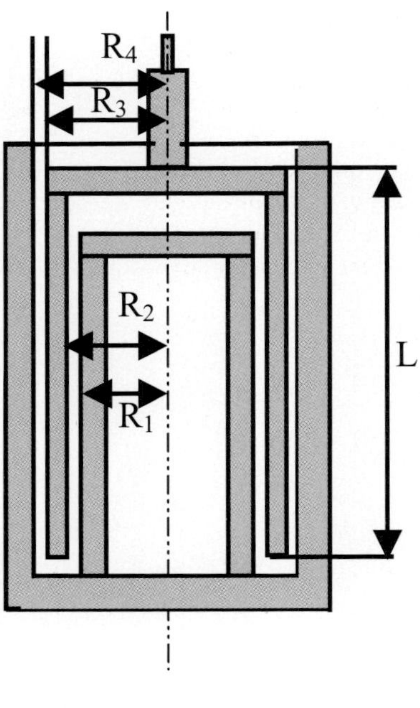

Fig. 5.12. Double gap system

According to DIN 53 453 the radius ratio d is

$$d = R_4/R_3 = R_2/R_1 \leq 1.15 \tag{5.18}$$

and the immersed length L

$$L \geq 3 \cdot R_3 \,. \tag{5.19}$$

This type of measuring system is obviously difficult to clean if it cannot be taken apart.

5.7 Flow Through Circular Capillary

So far we have considered the types of flow we are familiar with from cream application and mixing things like cake dough or a drink (stirring, not shaking). Long drinks are often served with a straw. Drinking a liquid through a straw is another type of flow, namely capillary flow. The straw can be thought of as a long tube, which we will now look at in more detail.

The fluid flowing through the tube (Fig. 5.13) adheres to the tube wall. As a result, a rate profile arises. The following applies at the tube wall:

$$r = R \quad \text{velocity} \quad v = 0 \tag{5.20}$$

and in the center of the tube:

$$r = 0 \quad \text{velocity} \quad v = \max \,. \tag{5.21}$$

Two forces are exerted on the liquid volume parallel to the tube axis:

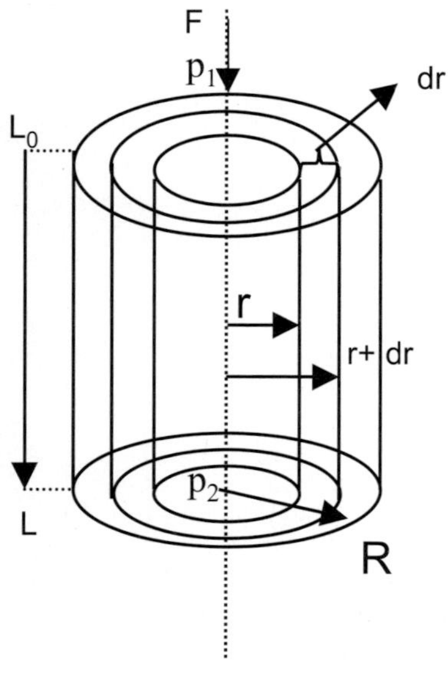

Fig. 5.13. Flow through a tube

1. The pressure force, which drives the liquid:

$$F_{\mathrm{d}} = r^2\pi(p_1 - p_2) \tag{5.22}$$

2. The frictional force:

$$F_{\mathrm{k}} = -\eta 2\pi r\,\mathrm{d}v/\mathrm{d}r\,; \quad \text{because} \quad \mathrm{d}v/\mathrm{d}r < 0 \tag{5.23}$$

The friction surface where the shear stress τ arises is in this case the cylinder area $2\pi r \cdot l$ of the flowing medium. For steady state flow the pressure and frictional forces must be opposite and equal:

$$F_{\mathrm{d}} = F_{\mathrm{r}} \tag{5.24}$$

$$r^2 \cdot \pi \cdot (p_1 - p_2) = -\eta \cdot 2\pi r \cdot L \cdot \mathrm{d}v/\mathrm{d}r \tag{5.25}$$

Solving this equation for the velocity derivative gives the following:

$$\mathrm{d}v/\mathrm{d}r = -1/2\eta \cdot (p_1 - p_2)\,/L \cdot r \tag{5.26}$$

Separation of variables and integration on the left from v to $v = 0$, and on the right from r to $r = R$ gives the following equation:

$$v(r) = 1/4\eta\,(p_1 - p_2)\,/L \cdot (R^2 - r^2) \tag{5.27}$$

This is the equation for a parabola where $y = 1 - x^2$. In other words, the velocity distribution is parabolic!

The next step is to calculate the total liquid volume flowing through the pipe per unit time. The liquid volume flowing through the zone $r + \mathrm{d}r$ per second is $\mathrm{d}Q$:

$$\mathrm{d}Q = 2\pi r \cdot \mathrm{d}r \cdot v(r) \tag{5.28}$$

Replacing $v(r)$ with the expression derived above [33] gives

$$\mathrm{d}Q = 2\pi r \cdot \mathrm{d}r \cdot 1/4\eta\,(p_1 - p_2)\,/L \cdot (R^2 - r^2) \tag{5.29}$$

Integration of the equation in the limits from $r = 0$ to $r = R$ gives the liquid volume flowing through the pipe per second:

$$Q = \int_0^R \mathrm{d}Q = \pi/8\eta\,(p_1 - p_2)\,/L \cdot R^4 \tag{5.30}$$

This is none other than the Hagen–Poiseuille law.

The volume flowing through a capillary per unit time also known as the flow velocity is proportional to the fourth power of the radius. This relationship discovered independently by both Hagen and Poiseuille was and is very important for the field of medicine. The capillary system of blood flow in humans has an approximate length of $L \approx 10^5$ km. An increase in muscular activity requires an

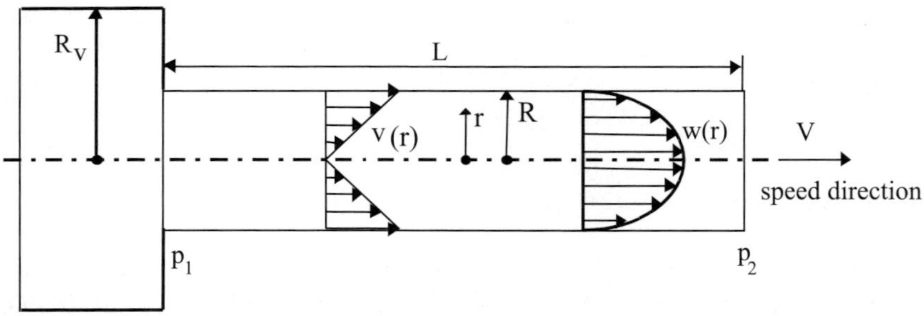

Fig. 5.14. Flow through a pipe – example of a circular capillary

increase in the velocity of blood flow Q. This is achieved by widening the capillaries because Q is approximately proportional to R. The increased demand for blood is met by reserves in the spleen and liver.

L = Length of capillary
R = Radius
Q = Volume flow or $Q = V/t$ volume per unit time
$\Delta P = P_2 - P_1$

For a circular capillary (Fig. 5.14) the viscosity can be obtained by rearrangement of the Hagen–Poiseuille Law:

$$\eta_{\text{app}} = \frac{\pi}{8} \cdot \frac{\Delta P \cdot R^4 \cdot t}{L \cdot V} \tag{5.31}$$

For the shear stress τ the following equation is obtained:

$$\tau(r) = \frac{F}{A} = \frac{\Delta P \cdot \pi \cdot R^2}{2 \cdot \pi \cdot R \cdot L} = \frac{\Delta P \cdot R}{2 \cdot L} \tag{5.32}$$

From this the shear rate for steady state laminar flow is derived:

$$\dot{\gamma}_{\text{app}} = \frac{4}{\pi \cdot R^3} \cdot \frac{V}{t} = \frac{4 \cdot \dot{V}}{\pi \cdot R^3} \tag{5.33}$$

We are speaking here also of the apparent shear rate, because the derivative was accomplished by a simplified assumption, as there are Newtonian behavior, stationary, laminar flow and incompressibility of the material. As in the PP system and the cylinder measuring system, the shear rate in the measuring gap is not constant and therefore must be corrected. This will be discussed in the next section.

5.8 Correction Methods

In the previous sections it was mentioned several times that for some measurement systems corrections [42] need to be made to determine accurately the flow properties of non-Newtonian fluids, as the shear rate shows non-linear behavior. Therefore consideration of the actual shear rate, which is not constant in the measuring gap, is very important. Not only the cone-plate model but all other models including the PP model, cylinder systems and circular capillaries need correction.

5.8.1 PP Measurement System

Rabinowitsch and Weissenberg were instrumental in obtaining the correct viscosity for non-Newtonian fluids in the PP model. They discovered that on the double logarithmic scale the corrected shear stress τ_c determined taking into account the slope $d\tau$ at $d\gamma$ gives a good approximation of the true value:

$$\tau_c = \frac{\tau_{\text{measured}}}{4} \cdot \left[3 + \frac{d \log \tau_{\text{measured}}}{d \log \gamma_R}\right] \tag{5.34}$$

5.8.2 Cylinder Measurement Systems

For the cylinder measurement systems the Schurz correction gives good results:

$$\gamma = \dot{\gamma}_{\text{measured}} \cdot \frac{1 - (R_i/R_a)^2 s}{s \cdot \left[1 - (R_i/R_a)^2\right]} \tag{5.35}$$

$$\text{where} \quad s = \frac{d \log \gamma}{d \log \tau} \tag{5.36}$$

5.8.3 Circular Capillaries

To determine accurately the true shear rate, corrections are needed. For correct working we have first to look for the right measurement. In Eqs. (5.35) and (5.36) we have to detect the flow loss pressure between intake and outtake of the capillary. However, we are not only measuring the pressure loss in the capillary but also the intake and outtake pressure loss. This loss can be corrected with a procedure of E.B. Bagley [43]. In addition different long capillaries are used to measure the pressure loss at the same speed. Plotting the pressure loss against the capillary length to diameter relationship you will get straight lines. Extrapolating this lines to the fictive capillary length of zero you get the so called Bagley pressure.

An apparent viscosity η_{app} is measured for non-Newtonian fluids because the viscosity is a function of the shear rate, which in turn is a function of the radius and hence variable.

$$\dot{\gamma} = \frac{\dot{\gamma}_{\max}}{4} \left[3 + \frac{d \log \dot{\gamma}_{\text{app}}}{d \log \tau_{\max}}\right] \tag{5.37}$$

The shear rate can be corrected according to Weissenberg and Rabinowitsch by stepwise calculation of the slope in the log vs log τ diagram, and the viscosities can be calculated with the corrected $\dot{\gamma}$ values. Because it takes a certain amount of time for laminar flow to evolve after the fluid enters the capillary, an entrance length LE can be defined in which the flow of layers near the wall is retarded and that near the axis accelerated:

$$\mathrm{LE} \approx 0.116 \;\; R - Re \tag{5.38}$$

The Hagenbach correction takes this into account and is especially important for short measuring times. A further correction is needed because friction is higher in the entrance zone. This can be recognized as an increased pressure drop. This error can be corrected according to Couette by an apparent lengthening of the capillary. Often the Hagenbach and Couette corrections are combined. Both effects are taken into account with an additive term.

In the following equation, m is a factor that must be determined using calibration oils or by measurements in two capillaries of the same diameter but different lengths:

$$\eta = \frac{\pi \cdot R^4 \cdot \Delta p \cdot t}{8 \cdot L \cdot V} - \frac{m \cdot \varrho \cdot \dot{V}}{8 \cdot \pi \cdot L} \tag{5.39}$$

It can also be seen that the additive term becomes very small with long measuring times, making it possible to eliminate the correction. If a non-Newtonian fluid with marked viscoelastic properties is measured, an additional pressure drop, as well as other effects, takes place because an energy-consuming elastic deformation takes place when the fluid enters the capillary. While in the capillary the fluid remains deformed, with relaxation occurring only upon exiting. This phenomenon is referred to as die swelling. At higher flow rates the melt can break. This will happen whenever the maximum elastic deformation of a specific sample is exceeded. Viscoelastic fluids exhibiting this behavior show a greater surface roughness with increasing shear rate. This must always be avoided because it prevents measurement of η and reduces the quality of the polymer. The elastic energy stored during die swelling is manifested as an apparent lengthening of the capillary.

5.9 Deformation and Relaxation

As we learned in Chap. 1, rheology is a branch of mechanics. So far we have considered the theoretical boundary conditions that are important for measurements. Now we will talk about the samples themselves. As a rule the sample to be studied will have several characteristic properties that depend largely on the type and magnitude of deformation [44] as well the time scale of deformation and observation. To understand this better it will help to consider a short experiment. When we press our finger against the center of a large window pane, the glass will bend a little. How much it bends will depend on the force we use. When we remove our finger, the pressure on the glass is removed and it returns to its original position. In other words, the glass is elastic. If we were to apply the same amount of pressure to the window pane for several years the glass would slowly start to flow or show viscous deformation. As a result, we would continue to see the bend even after the pressure is removed. The window pane will have been plastically deformed.

Therefore this sample shows different types of deformation for different deformation times. This example clearly illustrates the meaning of the terms plastic, viscous and elastic. Viscous deformation is the process of deformation, whereas plastic deformation is the end result of viscous deformation. Plasticity is another term for a material property. It represents the minimum force required to initiate deformation of a material. Figures 5.15 and 5.16 show mechanical bodies that can be described in terms of plasticity or elastic and viscous deformation.

In rheological measurements the phenomena of flow (deformation) and relaxation of fluid and solid systems are studied under applied external forces. Deformation is the relative displacement of material elements in which the cohesion of the model itself is not destroyed. Both solid and fluid materials are deformed when external forces are applied, but the way in which they are deformed and their response to subsequent removal of these forces differ. An ideal solid (Hookean model) releases the total input energy. It responds elastically and returns to its original shape after the energy is released. A spring is a model [45] for a Hookean model (Fig. 5.15). When the spring is loaded its deformation is proportional to the load. When the load is removed it returns to its original state.

An ideal fluid, on the other hand, starts to flow with the input of deformation energy. It shows viscous behavior. In this case the deformation is irreversible, for

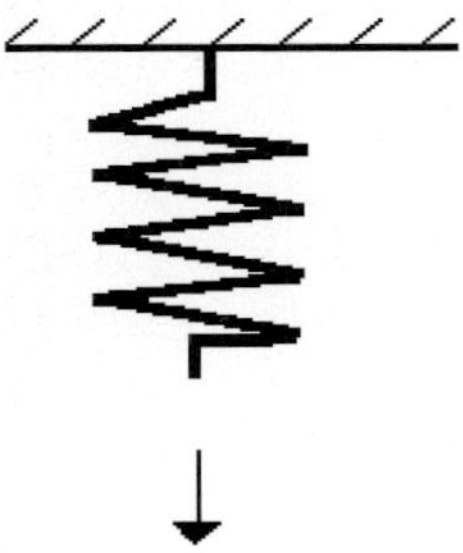

Fig. 5.15. Spring

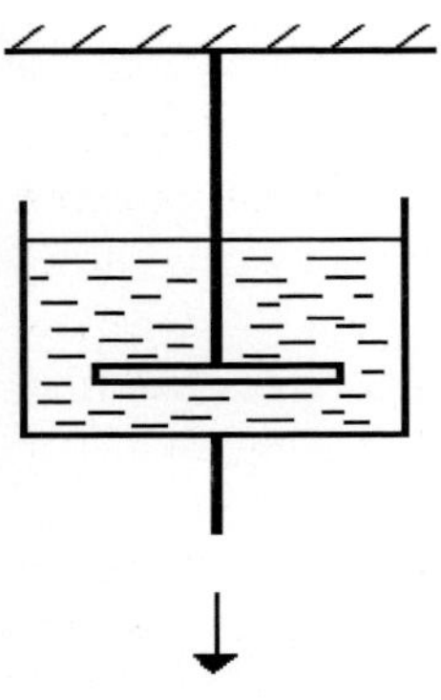

Fig. 5.16. Dashpot

the energy is completely converted to heat, making it impossible for the fluid to return to its original state on its own. The model for an ideal fluid is a dashpot. It consists of a cylinder filled with a liquid in which a piston is immersed that does not touch the walls of the cylinder (Fig. 5.16). For a fluid, the rate of deformation rather than the deformation is proportional to the applied force.

All real models show elastic as well as viscous responses to externally applied loads. After the load is removed, part of the added deformation energy is used to return to its original state and part is converted to heat energy and lost. Real material bodies are therefore said to be viscoelastic. Both components, the elastic and viscous, can be connected in series (Maxwell model) as well as in parallel (Voigt model, often also called the Kelvin model) (Figs. 5.17 and 5.18). Viscoelastic fluids are usually depicted as Maxwell bodies. Like a dashpot, which is connected in series, they do not fully return to their original state when the load is removed. The Voigt model serves as the model for viscoelastic solids. Connection in parallel allows complete recovery to the original state. However, this is an idealized model. Real substances only rarely show model behavior. Usually their behavior conforms to a complex combination of Maxwell and Voigt bodies.

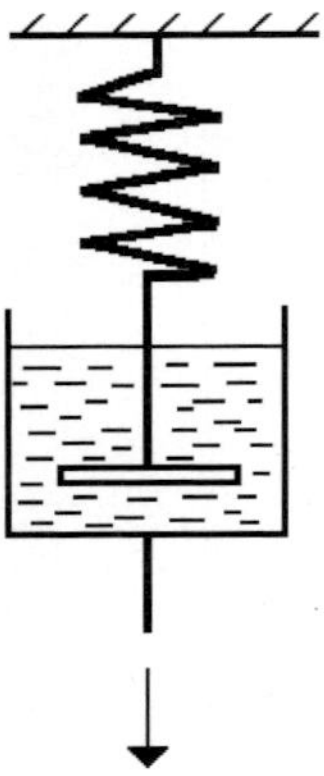

Fig. 5.17. Maxwell model

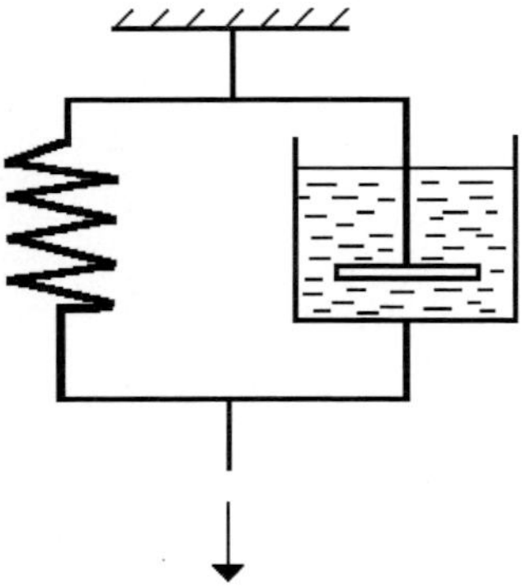

Fig. 5.18. Voigt model

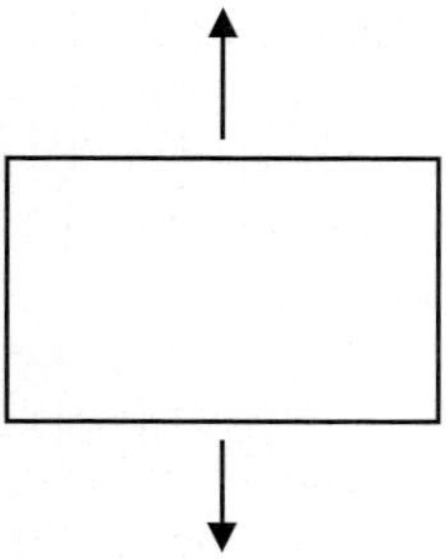

Fig. 5.19. Normal stress

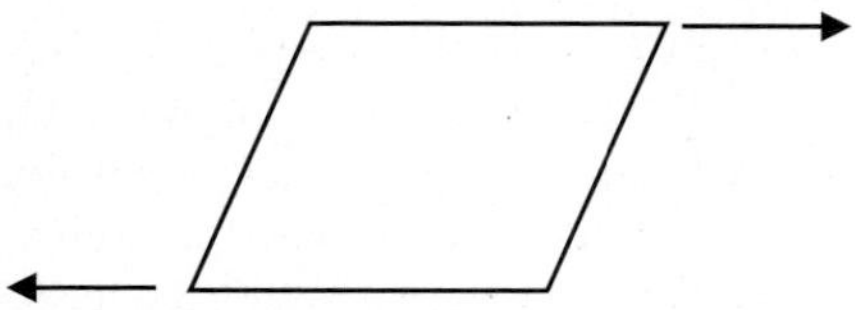

Fig. 5.20. Tangential stress

In mechanics, a distinction is made between certain types of forces. There are two basic types depending on the direction of loading. (1) Normal force where the load is applied perpendicular to the surface of the model (Fig. 5.19) and (2) tangential force where the load is applied parallel to the surface (Fig. 5.20). Normal stress (normal force/area) leads to elongation and compression, whereas tangential stress (tangential force/area) causes shear deformation. However, both types of stress can occur simultaneously as in bending.

5.10 Thixotropy and Rheopexy

The types of flow behavior discussed so far were a function of the shear rate. Thixotropy and rheopexy describe a flow behavior that is a function of shear time (Fig. 5.21, Fig. 5.22). At constant shear rate, the shear viscosity [46] of a thixotropic material decreases over time. During the shear time the bonding forces between molecules or particles diminish. Thinning of thixotropic substances is by definition reversible. In the subsequent resting state the network is regenerated by energy interactions and the original viscosity restored.

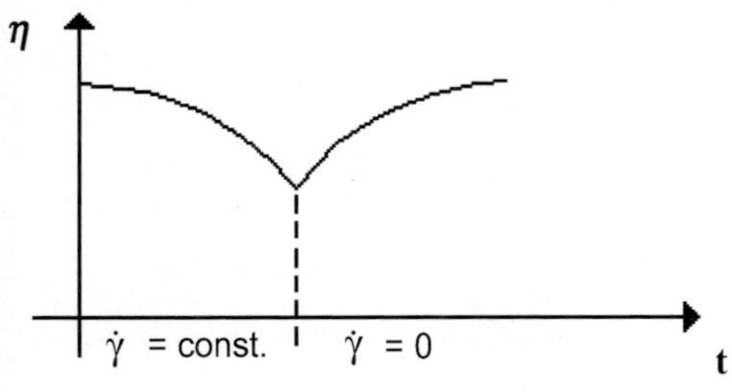

Fig. 5.21. Thixotropic flow

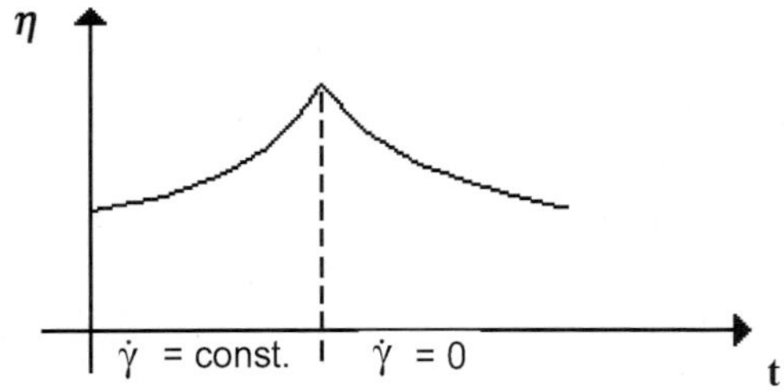

Fig. 5.22. Rheopexic flow

Rheopexy is the inverse of thixotropy. At constant shear rate the shear viscosity increases with shear time. This shear thickening is manifested as a viscosity increase. After a sufficient resting time the original viscosity is restored. True rheopexic flow behavior is very rare and should not be confused with gelation or hardening.

5.11 Vibration or Oscillation Measurements

So far, we have dealt only with uniform, accelerated movements. In nature, however, harmonic vibrations or sinusoidal movements [47] are common. Examples of harmonic vibrations include a tuning fork or the movement of a pendulum in a place-time diagram.

A tuning fork looks like a two-pronged fork in the shape of an elongated U. When tapped, the prongs vibrate back and forth, producing nearly a sinus tone. The pitch of the tone depends on the length and mass of the vibrating prongs (required accuracy $\pm 0.5\,\text{Hz}$). The tuning fork is thought to have been invented in 1711 by the English musician John Shore. A sinusoidal vibration (apart from dissipation of the vibration = damping) can be visualized (Fig. 5.23) by attaching a paint brush to one of the prongs vibrating in the y direction. The brush traces the movements in the y direction on a strip of endless paper pulled underneath it at a constant speed in the x direction.

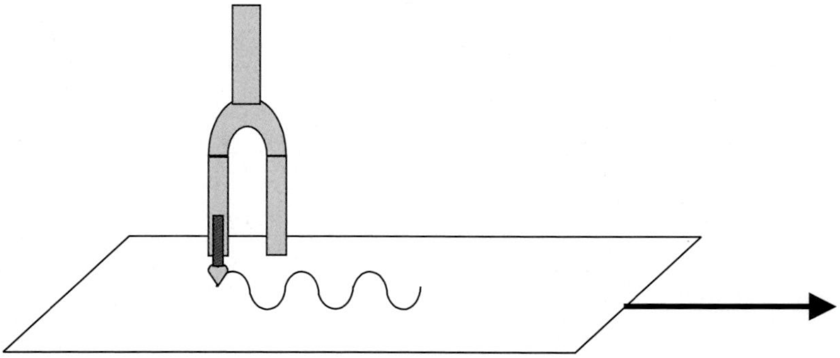

Fig. 5.23. Visualization of a sinusoidal vibration with a tuning fork

5.11.1 Steady and Dynamic Stress

In steady measurements the material is deformed by continuous rotation. The cylinder, plate or cone rotates in one direction at constant or variable shear stress or shear rate. The dynamic viscosity is calculated from the measured velocity or torque and the preset shear stress or strain.

In dynamic measurements the material is subjected to an oscillating shear stress or strain. Because movement of the rotating system is sinusoidal, we speak of an oscillation measurement. Whether the shear stress or strain oscillates is irrelevant.

For deformation from a defined strain, the amplitude (γ) is preset in a dynamic measurement (Fig. 5.24). The frequency chosen determines the period of the oscillation. Since this is an angular frequency in rotational viscometers, the frequency is expressed as 1/s or rad/s: $\omega = 2\pi f$.

The sinusoidal strain can be described mathematically as a harmonic oscillation by the distance-time law:

$$\gamma = \widehat{\gamma} \sin(\omega t) \tag{5.40}$$

Deformation of an ideal elastic model is proportional to the load. Consequently, the force (shear stress) [48] is greatest at maximum deformation. If there is no deformation, the force is zero. The stress and strain curves will be in phase for a sinusoidal load.

An ideal viscous fluid is characterized by a proportionality of the force (shear stress) and shear rate (Fig. 5.25). Maximum shear rates correspond to maximum forces. Derivation of the strain over time gives the shear rate:

$$\dot{\gamma} = \frac{\mathrm{d}\gamma}{\mathrm{d}t} = \omega\widehat{\gamma}\cos(\omega t) = \widehat{\dot{\gamma}}\sin\left[(\omega t) + \frac{\pi}{2}\right] \tag{5.41}$$

There is a phase lag of 90° between the shear rate and the strain. At maximum strain the shear rate is zero, and at the point of inflection of strain the shear rate

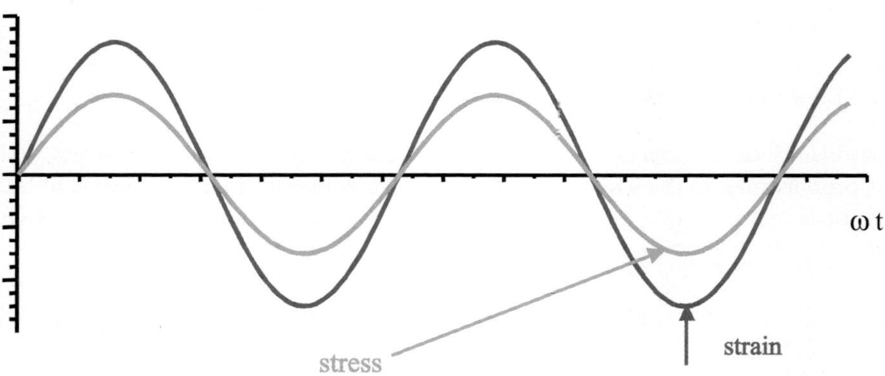

Fig. 5.24. Strain and stress curves for an ideal elastic solid (Hookean solid) with dynamic strain

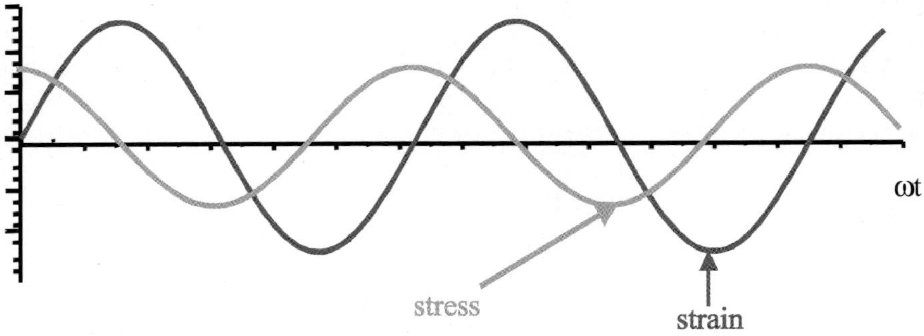

Fig. 5.25. Strain, shear rate and shear stress curves for an ideal fluid with dynamic strain

reaches its maximum. The product of angular frequency and shear amplitude is the shear rate amplitude γ.

Real materials exhibit linear viscoelastic behavior at small amplitudes. For a transient sinusoidal shear strain an oscillating out-of-phase shear stress arises with the amplitude τ:

$$\tau = \widehat{\tau} \sin\left[(\omega t) + \delta\right] \tag{5.42}$$

The phase angle δ gives the shift in the response oscillation on the time axis.

5.11.2 Ideal Elastic Solids

The shear stress curve for an ideal solid is sinusoidal and in phase with the strain curve. The material-specific shear modulus is the quotient τ/γ. Therefore the shear stress is

$$\tau = G_{\mathrm{E}}\gamma = G_{\mathrm{E}}\widehat{\gamma}\sin(\omega t) \tag{5.43}$$

Since the shear stress and strain are proportional, G_{E} is a constant.

5.11.3 Ideal Viscous Fluids

An ideal fluid responds to a sinusoidally applied strain with a shear stress that is proportional to the shear rate. The phase lag between the shear stress and the strain is $\pi/2$. According to Newton, $\eta_{\mathrm{v}} = \tau/\dot{\gamma}$ and hence the shear stress is expressed by

$$\tau = \eta_{\mathrm{v}}\dot{\gamma} = \eta_{\mathrm{v}}\omega\widehat{\gamma}\sin\left[(\omega\tau) + \frac{\pi}{2}\right] = \widehat{\tau}\sin\left[(\omega\tau) + \frac{\pi}{2}\right] \tag{5.44}$$

The product of the viscosity, angular frequency and shear amplitude is the shear stress amplitude τ.

5.11.4 Real Solids

Real solids have elastic as well as viscous components. Such viscoelastic materials respond to an applied strain like a Voigt model (Fig. 5.18). Because the two components are in series, the resulting shear stress is

$$\tau = G_E\widehat{\gamma}\sin(\omega t) + \eta_V\omega\widehat{\gamma}\cos(\omega t) \tag{5.45}$$
$$\tau = \widehat{\gamma}\left[G_E\sin(\omega t) + \eta_V\omega\cos(\omega t)\right] \tag{5.46}$$

The shear stress can also be expressed as follows:

$$\tau = \widehat{\gamma}\left|G^*\right|\left[\sin(\omega t) + \delta\right] \tag{5.47}$$

where

$$\left|G^*\right|_V = \sqrt{G_E^2 + (\omega\eta_V)^2} \tag{5.48}$$

and

$$\tan\delta = \frac{\omega\eta_V}{G_E} \tag{5.49}$$

The shear stress curve is likewise sinusoidal but the phase lag of the strain curve is the phase angle δ. It is proportional to the shear amplitude and the contribution of the complex modulus G^*. The response of a Voigt model to dynamic (oscillating) strain is by definition independent of the frequency.

However, the parameters of viscoelastic materials are usually frequency dependent. The frequency dependent moduli $G'(\omega)$ and $G''(\omega)$ are also known as the storage and loss module respectively. They describe the elastic and viscous components of a material. The viscous component G'' is obtained from the product of the dynamic viscosity $\eta'(\omega)$ and the angular frequency ω. The shear stress is:

$$\tau = \widehat{\gamma}\left[G'(\omega)\sin(\omega t) + G''(\omega)\cos(\omega t)\right] \tag{5.50}$$
$$\tau = \widehat{\gamma}\left|G^*(\omega)\right|\sin\left[(\omega t) + \delta(\omega)\right]\ . \tag{5.51}$$

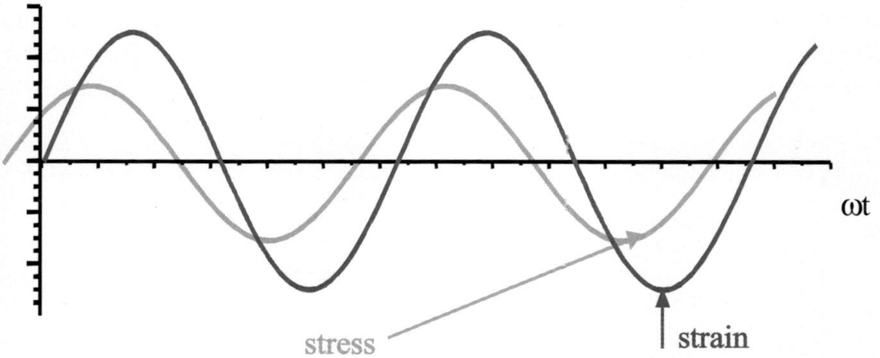

Fig. 5.26. Strain and shear stress curves for a viscoelastic solid (Voigt model) with dynamic strain

Accordingly,

$$|G^*| = \sqrt{G'^2 + G''^2} \tag{5.52}$$

and

$$\delta = \arctan \frac{G''}{G'} \tag{5.53}$$

5.11.5 Complex Representation

The oscillatory experiment can also be described using complex numbers. The complex number G^* is an ordered pair of real numbers where G' is the real part and G'' the imaginary part (Fig. 5.27). The mathematical formula is:

$$|G^*| = [G'; G''] = G' + i^* G'' \tag{5.54}$$

In the plot, the real part G' is projected on the x-axis (real axis) and the imaginary part G'' on the y-axis (imaginary axis).

Euler's formula $e^{i\varphi} = \cos\varphi + i\sin\varphi$ converts the trigonometric form of complex numbers to the exponential form. Consequently, the shear stress and strain can be expressed as follows:

$$\tau = \dot{\gamma} G^* = \widehat{\gamma} G^* e^{i\omega t} = \widehat{\tau} e^{i\omega t} \tag{5.55}$$

$$\gamma = \widehat{\gamma} e^{i\omega t} \tag{5.56}$$

This gives the following relationships for the complex modulus G^*:

$$G^* = \frac{\tau}{\gamma} = \frac{\tau}{\gamma} = \frac{\widehat{\tau}}{\widehat{\gamma}} = G' + iG'' \tag{5.57}$$

Since the strain of a viscoelastic material lags behind the shear stress by the phase angle δ, the following vectorial representation results:

The length of the vector corresponds to the amplitude of the shear stress (τ) or the strain (γ); the projection on the real axis at the time t gives the momentary shear stress or strain value (τ and γ) at that frequency. In the diagram (Fig. 5.28) $G'\gamma$ is the elastic component that is in phase with the strain. The viscous component

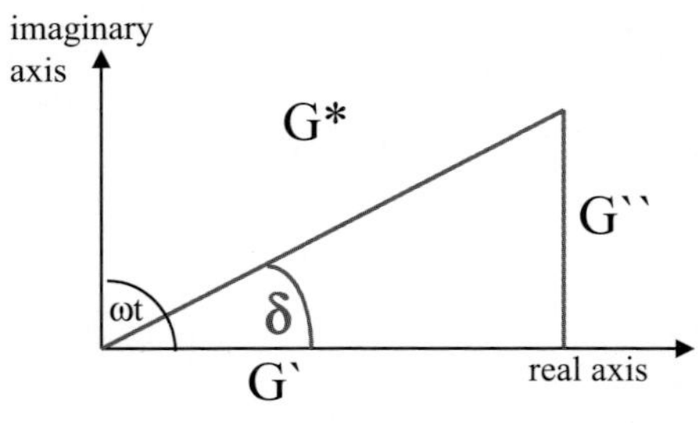

Fig. 5.27. Plot of the complex modulus G^*

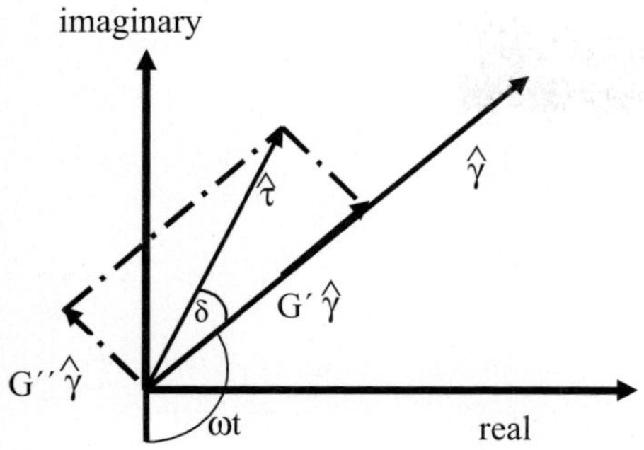

Fig. 5.28. Complex vectorial representation of dynamic stress

$G''\gamma$ runs ahead of the strain and is therefore out of phase. The shear stress rotates counterclockwise at an angular velocity of ω.

The derivative of strain gives the shear rate:

$$\dot{\gamma} = \widehat{\gamma}\mathrm{i}\omega\,\mathrm{e}^{\mathrm{i}\omega t} \tag{5.58}$$

The complex dynamic viscosity can be calculated analogously from the complex modulus:

$$\eta^* = \frac{\tau}{\dot{\gamma}} = \frac{G^*}{\mathrm{i}\omega} = \frac{G''}{\omega} - \frac{G'}{\omega} = \eta' - \mathrm{i}\eta'' \tag{5.59}$$

Therefore, the complex dynamic viscosity can be seen as the difference between η' and the imaginary part η''.

Important for normal use is the meaning of:

G' = storage modulus ⇒ elastic behavior;
G'' = loss modulus ⇒ viscose behavior;
$\tan\delta$ = loss factor ⇒ quotient: G'' divided by G'; and
η^* = complex dynamic viscosity.

6 Measuring Instruments

In today's fast-paced world it is more important than ever to be the first on the market with new products. One way to achieve this is to shorten development times. To compensate for the risks this involves attempts are underway to make greater use of the potential of modern analytic techniques. But even basic research projects would be unthinkable today without the expertise derived from analytics. Rheology plays an important role in analytics because the insights obtained from rheological findings are of fundamental importance to materials science and process engineering and hence to quality control as well as to research and development. For instance, viscosity data are essential for specification of raw materials, intermediate and finished products, for monitoring and regulating the production process but also for developing new production processes. The type of viscometer or rheometer used to obtain this data depends on the demands placed on the measuring results. How accurate does the viscosity value need to be? What will the result be used for – process control or a general statement on the structure of the sample? Or would a viscosity curve be a better alternative? Obviously, deciding which measuring instrument to use is not an easy task. What factors should determine the instrument chosen? Some of the many questions that need to be answered in order to select the right instrument are:

- Should it be temperature-controlled?
- Is a single point measurement sufficient or is a flow curve needed?
- If a flow curve is needed, the next question is which shear rate range to use – low or high?
- Are stationary measurements adequate or would oscillation measurements be more informative?
- Should the instrument provide "relative" or "absolute" values?
- Would a shear stress or a speed control instrument be preferable?

To answer these questions sufficient measuring data and experience with the samples to be measured and the available instruments are needed. Since viscometers and rheometers can be very expensive and also, depending on the requirements, need trained personnel this type of investment should be considered very carefully. If neither an instrument nor experience is available, instrument manufacturers can be helpful. In their applications laboratories one can perform measurements or have samples measured. It is also possible to rent an instrument for a small fee.

Also useful are contacts to engineering schools and universities. A Master's thesis can be a good way to obtain the basic information needed for making a decision on this type of investment.

If trained personnel as well as adequate funding are available, the next step is to select the right instrument. Due to an important physical relationship between force and distance, only two types of measuring instruments can be constructed from a physical viewpoint. Either the force is defined in an experiment and the resulting distance is measured or the distance is defined and the force is measured as the response signal. Consequently, the instruments must be either shear stress or shear rate controlled.

6.1 Modern Rheometer

From modern rheological analytical data it should be possible to determine the flow properties of materials and the relationships between their structural properties as a function of temperature (−50 to +100 °C) and an applied force ranging from less than a few 0.001 mNm up to 10 Nm (or in the shear rate range from 0.00001 up to 100 000 s^{-1}). Unfortunately this means that one instrument will never be able to do every measurement. A second rheometer will always be needed for high shear rates, as one rheometer can never cover the whole measuring range.

Shear stress controlled instruments come as torsional or rotational viscometers and high performance capillary rheometers and shear rate controlled instruments as rotational viscometers.

In a rotational viscometer the test sample is placed between two symmetrical rotating bodies. Mainly plate/plate, plate/cone and coaxial cylinder measuring systems are used (Fig. 6.1). The force leading to deformation of the sample is defined by the applied torque. Since the test substance exerts a certain resistance to rotation that depends on the viscosity, the measuring probe rotates at a certain speed. The shear rate is calculated from the frequency ($f = \omega/2\pi$) and the geometry

Fig. 6.1. Shear stress controlled rheometer with cylinder system

of the measuring system. The viscosity is obtained from the ratio of the shear stress to the shear rate.

Rotational viscometers are either of the Searle or Couette type. In coaxial cylinder measuring systems of the Searle type the inner cylinder rotates while the outer cylinder does not. The torque sensor is also attached to the inner cylinder. In cone/plate and plate/plate measuring systems the cone or top plate rotates. The measuring probe used depends on the type of system and here is also connected to the rotating body. The temperature is usually controlled by a Peltier element (Fig. 6.2.) which is based on a thermoelectric effect named after the French physicist [49] Charles Athanase Peltier. Electrical power is converted directly into thermal output. At the junction between two materials of different conductivity, heat either evolves or is absorbed when electric current flows through the circuit.

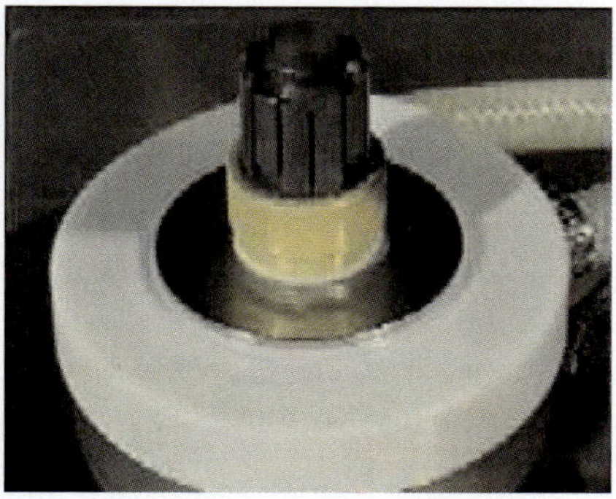

Fig. 6.2. P/P with Peltier element

High cooling and heating rates can be achieved with a Peltier element. The heat generated must be removed via a water bath. Because unfortunately only the bottom plate is cooled, a temperature gradient builds up in the sample that can be significant depending on the thickness of the sample.

This type of instrument is offered by many manufacturers [50, 54] including Bohlin, GABO Qualimeter, Physica-Meßtechnik, ThermoHaake and TA Instruments.

They differ mainly in their appearance but have very similar technical specifications. There are also differences in the software and in the extensive analytical models.

If the drive and measuring probe in a torsional or rotational viscometer are decoupled the instrument is said to be a speed or shear rate controlled rheometer. The bottom part of the measuring system (plate or outer cylinder) rotates at a defined speed. The energy introduced in this way is transferred by the sample to be measured to the top part of the measuring system (cone, plate of inner cylinder) and the resulting shear stress calculated from the measured torque. The ARES instrument from Rheometrics-Linie, a subsidiary of TA Instruments, is one of the best instruments on the market, but the other manufacturers named above also offer rate-controlled instruments. Another advantage of the instrument is a very

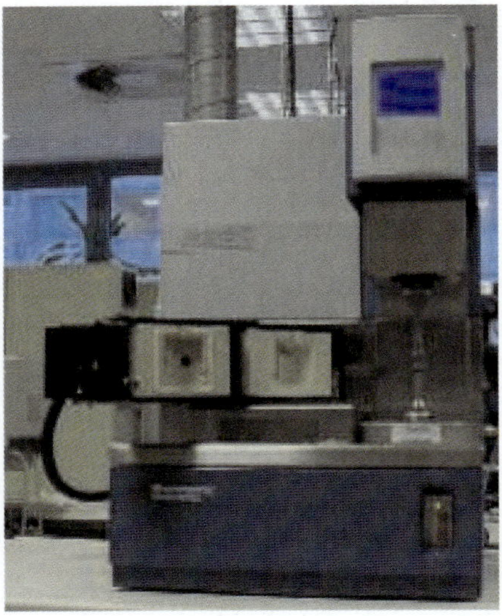

Fig. 6.3. ARES rheometer

precise temperature control via an oven. Cooled or heated air flows around the entire sample. The temperature gradient measured inside the oven is less than $\pm 0.5\,^\circ\mathrm{C}$.

The two types of rotational rheometer describe here so far have another thing in common; they can be used for any stationary, dynamic or oscillation, measurement!

- The temperature range depends on the heating and cooling units used and extends from $-150\,^\circ\mathrm{C}$ to not more then $500\,^\circ\mathrm{C}$ with liquid nitrogen and an electric heater or from -50 to $150\,^\circ\mathrm{C}$ with a Peltier element.
- The shear rate of course depends on the sample and ranges from $0.00001\,\mathrm{s}^{-1}$ to not more than $1000\,\mathrm{s}^{-1}$.
- The dynamic frequency ranges from 0.0001 to 500 rad/s.
- The shear stress range starts at 0.01 mNm and ends at 200 Nm maximum.

6.2 High Shear Rheometer

Even higher shear rates like those used to measure the viscosity while a product is being pumped are achieved with a high performance capillary viscometer. The Rheograph 2002 [55] from Göttfert (Fig. 6.4) is this type of instrument. It allows adjustment or premeasurement of production conditions in the temperature range from 25 to $300\,^\circ\mathrm{C}$ and in the shear rate range from 100 to $250\,000\,\mathrm{s}^{-1}$. Pressures up to 1500 Pa can arise.

The minimum requirements for measuring over the whole shear rate range from 0.00001 to $250\,000\,\mathrm{s}^{-1}$ are a shear stress-controlled instrument and a high pressure capillary rheometer. An even better solution can be to use the results from a rate-controlled instrument.

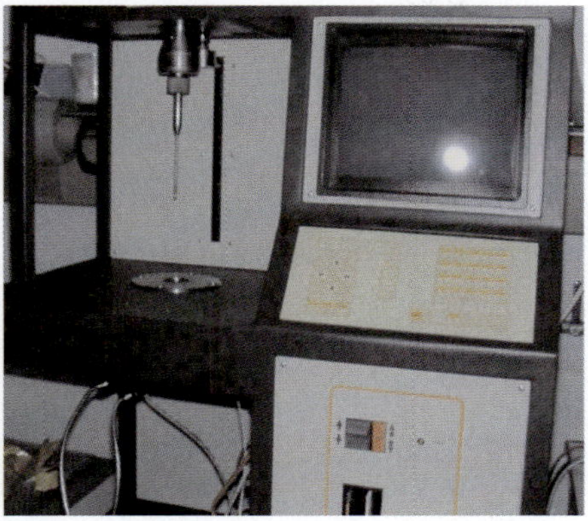

Fig. 6.4. High shear capillary rheometer

6.3 Standard Viscometer

With these measuring results the rheologist is now able to determine the correct boundary conditions for every sample and problem. In many cases this means, however, that once the measuring conditions are adapted to the problem, it is determined that the measurement can be performed with less expensive instruments [56–58] such as the Brookfield (Fig. 6.5), Rheotest, Coesfeld and formerly Contraves (Fig. 6.6) instruments.

With these instruments flow curves can be measured in a limited shear rate range from about 1 to $100\,s^{-1}$. The temperature must be recorded simultaneously. This means they can be used optimally for control of incoming goods but also in production, where comparative measurements can be used to monitor conformance to required viscosity data at constant temperature and defined shear gradient.

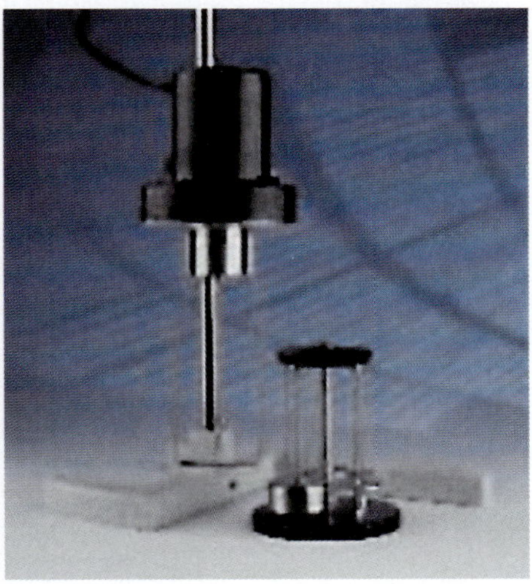

Fig. 6.5. Brookfield

Fig. 6.6. Rheomat 180

6.4 Often Used Viscometer

If the samples to be measured are Newtonian fluids, even more economical instruments such as the flow cup (Fig. 6.7), Ubbelohde or Vogel–Ossag (Fig. 6.8) instruments can be used, as only a relatively constant temperature needs to be ensured. This can be achieved by performing the measurements in an air conditioned room.

Temperature-controlled single-point instruments like the falling ball viscometer (Fig. 6.9) are also available. Many of these instruments were constructed for the first time 20 years ago but are still in use today in places where it is important to measure changes or deviations from a required value. Because of the different shear rates achieved with different instruments it is entirely possible to find three or more of these single-point instruments in use in quality assurance. Modern rheometer are a good alternative because the widest variety of shear rates can be achieved at a constant temperature with one instrument.

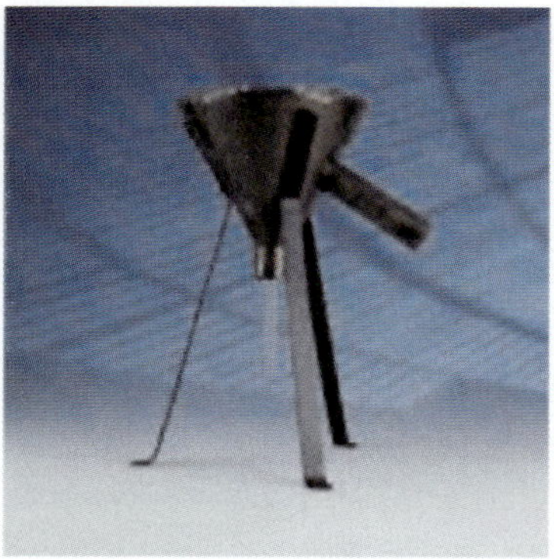

Fig. 6.7. Flow cup viscometer

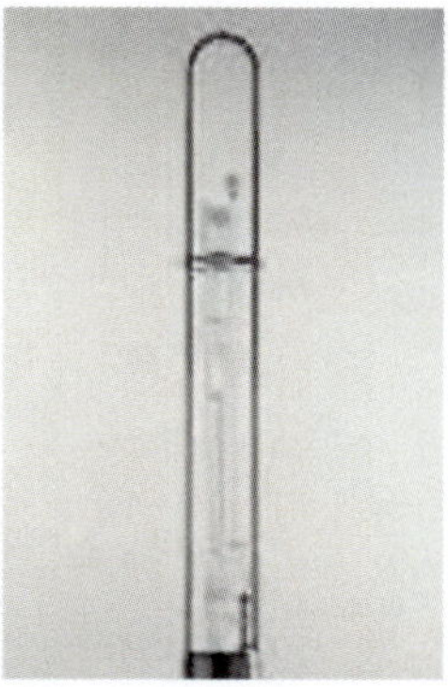

Fig. 6.8. Vogel–Ossag

Fig. 6.9. Falling ball

6.5 Automatic Sampler

If many similar samples need to be measured, a rheometer with automatic sampler like that from Physica Messtechnik [52] can be used. Up to 32 samples (Figs. 6.10 and 6.11) can be measured consecutively at a very constant temperature. The measurements are fully automated as even cleaning of the measuring cylinder after each measurement is done automatically.

Fig. 6.10. Rheometer with sample holder

Fig. 6.11. Sample carousel

6.6 In-process In-/On-line Viscosity Measurements

There are three types of in-process viscometers:

1. Off-line rheometers, which include all laboratory instruments
2. On-line rheometers, which are usually operated in the bypass mode
3. In-line rheometers, which measure directly in the processing vessel such as a mixer

For off-line rheometer measurements, a sample is taken from the running production process and subsequently measured in the laboratory. The disadvantage is obvious: it takes a very long time to obtain a result. The other two types are much more suitable for process control because a measuring result is available in a relatively short time. Torsion rods, ultrasound technology, dielectric measurements and capillaries are used. Measuring the viscosity in a mixing tank is difficult because the sample volume to be measured is not at rest during the measurement and therefore the speed and temperature of the sample also need to be measured separately, as they affect the actual measurements.

Torsion rods come in many shapes (Fig. 6.12) and designs. Depending on their geometry, they oscillate at frequencies ranging from approximately 2800 to 6800 Hz. This type of rod is made to oscillate [59] at a constant frequency. If it is

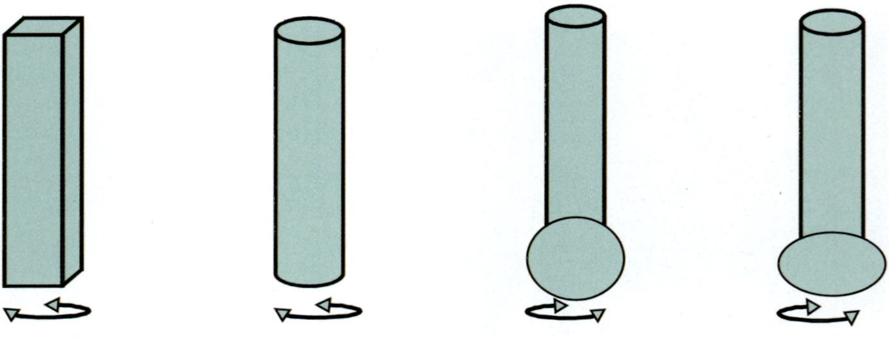

Fig. 6.12. Working principle of torsion rods

immersed in a fluid, a damping of the oscillation takes place that depends on the fluid. The viscosity can be calculated from this damping. Torsion rods can be used in- as well as on-line.

Other techniques that are more suitable for use on-line, i.e. built into a bypass, include laser Doppler, ultrasonic and dielectric measurements.

The laser Doppler and ultrasonic (Fig. 6.13) techniques work according to a similar principle [60], but in the first case light and in the second sound waves are transmitted from one side of the sample and either received at the other side or the runtime of the waves determined by reflectance. In these measuring techniques use is made of the direct correlation of the density differences between the sample to be measured and the surrounding air or interface and the differences in transit times in the separate media. Often these measuring methods have natural limits. For instance the laser beam cannot penetrate through a non-transparent fluid or the ultrasound method cannot be used if the cross-section is too large.

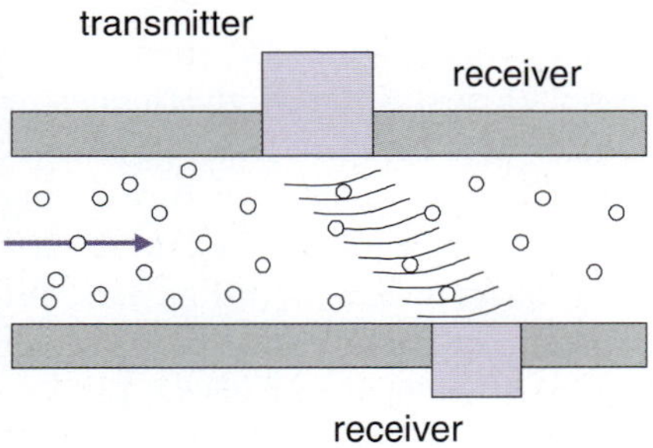

Fig. 6.13. Ultrasonic system

Dielectric spectroscopy (Fig. 6.14) is another very interesting test method [61]. It is known that cosmetic emulsions contain water and that water consists of dipoles. Depending on the applied frequency these dipoles align themselves in an alternating electric field more or less closely with the network structure of the emulsion to be measured. The frequency range for this type of measuring system is from less than approximately 20 Hz to several 1.5 MHz.

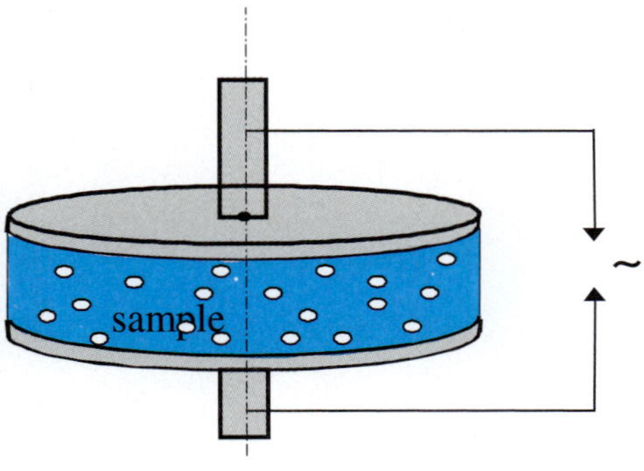

Fig. 6.14. Dielectric spectroscopy

Careful consideration must be given as to where to set up this type of instrument in the production process for measurement of the viscosity. It must of course be easily accessible for maintenance and cleaning. It should allow non-contact operation, flexible measurements in a large shear stress range, accurate temperature measurement of a relatively large undiluted sample volume and also not be too expensive.

Use of capillaries is an inexpensive alternative. Let us recall the two equations for calculation of the shear rate and shear stress:

$$\dot{\gamma} = \frac{4}{\pi R^3} \cdot \dot{V} \tag{6.1}$$

$$\tau = \frac{\Delta P \cdot R}{2L} \tag{6.2}$$

It is striking that the shear rate does not depend on the length L if the diameter D is the same, but if the length L is the same and the diameters D_1 and D_2 differ, different shear rates result.

By using a double (Fig. 6.15) or triplet capillary system (Fig. 6.16), the viscosity can be measured simultaneously in a bypass with two different shear rates or information obtained on the viscoelastic behavior with the same shear gradient.

Much work still needs to be done before it will be possible to measure during the production of an emulsion the really important parameters that can be used to regulate the process. There are many ways to collect on-line or in-line data, but the critical information on viscosity changes during production and their effect on the intermediate product and hence on the finished product is not always known.

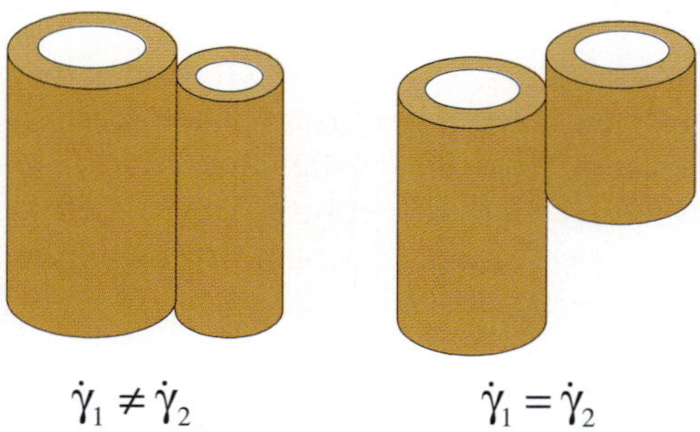

Fig. 6.15. Double capillaries for use in an on-line rheometer

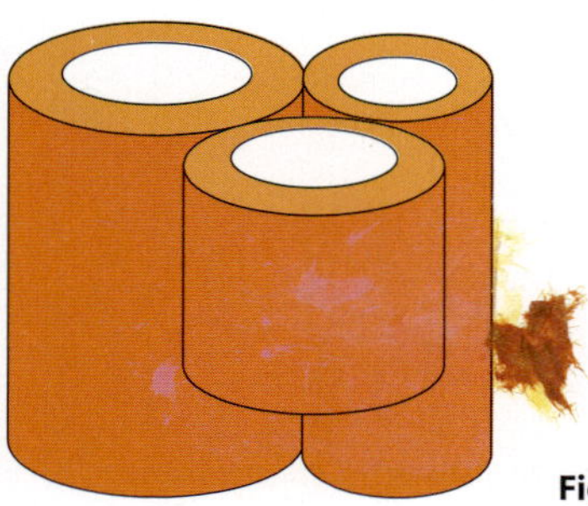

Fig. 6.16. Triplet capillary system

6.7 Future Prospects

Determining the stability or more precisely the storage life of a product is one of the most important tasks in R&D and one in which the rheologist plays an important role. Physically, it is impossible to extrapolate from measurements taken in a limited time frame how a product will change in the future. One solution to this dilemma is to attempt to simulate an accelerated aging of samples by applying an external force. This might be a higher temperature or even a deformation. By directly comparing the results of such stress tests it is possible to say whether the stability of a sample is better than, the same as, or worse than a reference sample.

A more scientific formulation of the question would be: what happens to an emulsion when it ages? The signs of aging that everyone can recognize such as a color change or separation of water or oil indicate that an emulsion is no longer stable. It takes up to three years to do this type of storage test, which is routinely performed in development laboratories. Visible changes in emulsions can only arise if the internal structure of an emulsion changes. An attempt can be made to measure this structural change, and this is where the rheologist comes in.

Fig. 6.17. Rheometer with UV tool

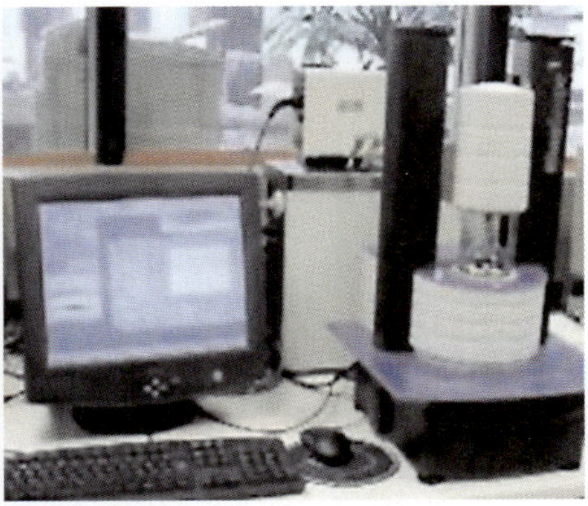

Fig. 6.18. RheoScope 1

Currently, institutions and universities like MIT are already working on the construction of a new instrument generation, the microrheometer. Soon nanorheology will become a reality. Instrument manufacturers and users from universities and industry are currently concentrating their efforts on instrument combinations such as linking rheometer to DSC, UV detectors (Fig. 6.17), dielectric measurements or optic systems like microscopes and video cameras.

The human eye will always be able to process a picture better than a measuring curve or even a data table. This is why the Rheoscope 1 from Thermo-Haake, which combines an optical instrument with rheological measurement was a logical development. As always, however, the question remains whether the new findings obtainable by combining instrumental techniques justify the higher costs and whether there will be a return on investment for the instrument user.

7 Most Important Test Methods

A variety of measurement methods and instruments are used for rheological characterization of cosmetic emulsions. At low shear rates [62] for detection of the apparent yield stress, stressed-controlled rheometer are used because with this type of instrument the stress is preset and the onset of flow can be determined more or less accurately depending on the quality of the angle resolution of the instrument. Characterization of processing conditions requires measurements at high shear rates with a torsional rheometer or at extremely high shear gradients with a capillary viscometer. The measurement system used is also important and will depend on the product to be measured. Watery fluids need a cylinder system, viscous samples the cone-plate system and all other solid materials the parallel plate system. Modern rheometers are versatile and allow different types of strains and combinations thereof, permitting simulation of many operations in a production process.

One of the first experiments performed with a new, unknown sample is the so-called positive ramp test (Fig. 7.1). Here the applied shear stress is increased continuously with time at a constant temperature. It is also referred to as a stress-time ramp. In this experiment an integral viscosity is obtained as a first estimation of the range in which the actual measurements should be performed. A step test is then performed as the second experiment (Fig. 7.2). In contrast to the first

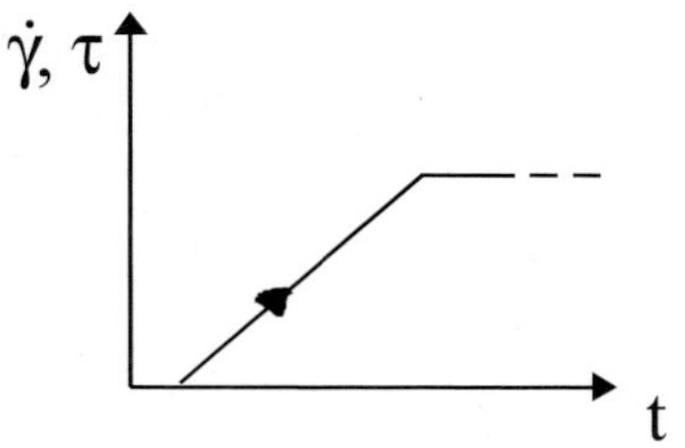

Fig. 7.1. Positive ramp

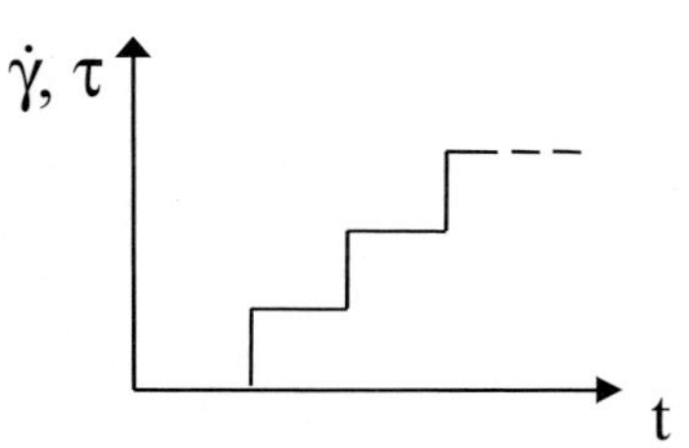

Fig. 7.2. Step test

experiment, here the shear stress is kept constant for a certain time and then increased in several steps. The result obtained is a so-called static viscosity value.

Next the material properties are determined after sudden application (Fig. 7.3) or removal (Fig. 7.4) of a load.

After determining the integral viscosity in a positive ramp test as shown in Fig. 7.1 it is of course also possible to do a negative ramp test (Fig. 7.5) to see whether the sample had been changed by the original positive ramp test. Last but not least, the user is free to choose any combination of these tests (Fig. 7.6) and even change temperatures.

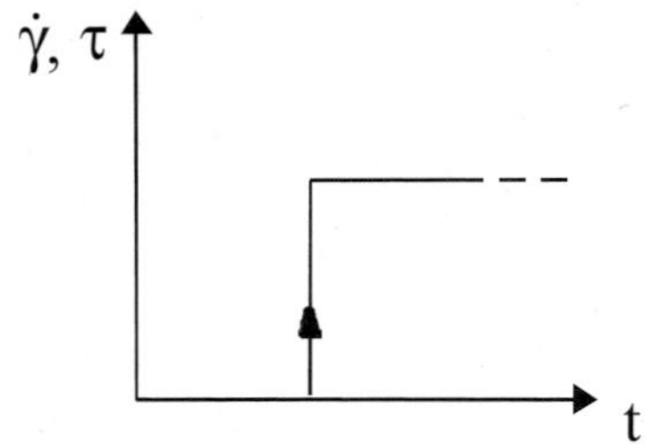

Fig. 7.3. Load jump

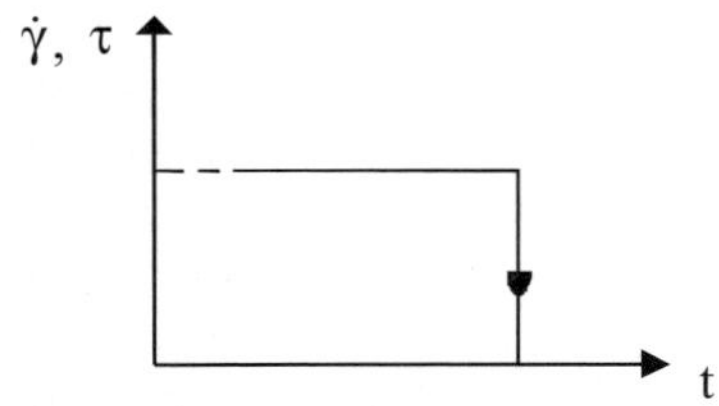

Fig. 7.4. Release jump

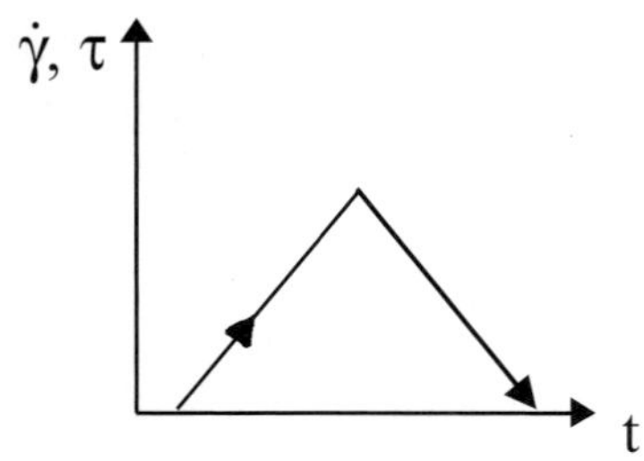

Fig. 7.5. Positive and negative ramp

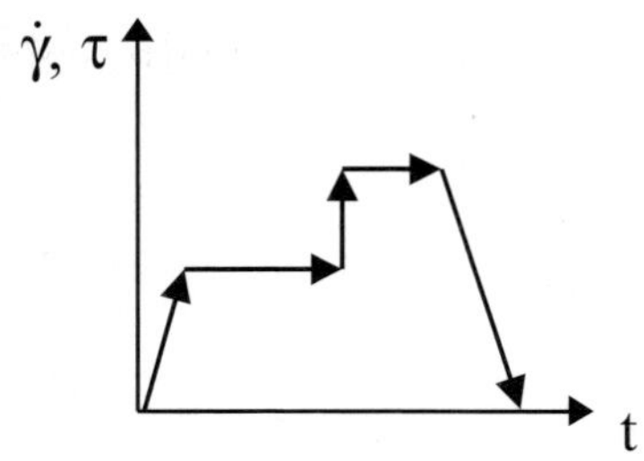

Fig. 7.6. Combination test

7.1 Stress Ramp Test

Now let us take a closer look at the positive ramp or stress ramp test. It is a simple and quick test that can be done with a stress-controlled rheometer. The shear stress is increased continuously within a certain time (Fig. 7.7), the resulting torque measured at discrete time intervals and from this the viscosity calculated. Programming the right stress ramp is crucial. But how can we determine the right ramp? Should it be 20 Pa/min or 100 Pa/min? The experiment provides the answer.

Let us assume that the sample has a yield stress also known as the yield point. This is product-specific and should be determined relatively accurately. A small stress ramp should be programmed for thinner liquids and a larger one for thicker creams. An appropriate number of measurements must be performed to determine if the whole product range can be measured with a single stress ramp.

After programming the stress ramp and performing the measurement the next step is to interpret the resulting curve. Although there are several models available for interpreting measuring curves, they often give rise to relatively large errors because ideal mathematical conditions usually do not exist. Differences already arise from the way in which the results are plotted. If the shear stress is plotted against the shear rate, the curve shown in Fig. 7.8 is obtained. Initially, the force or shear stress applied to the sample does not cause any detectable deformation. In other words, the system remains at rest.

No deformation becomes visible until a critical shear stress has been reached, and a shear rate is determined. Using the mathematical relationship formulated by Herschel–Bulkley:

$$\tau = \tau_0 + K \cdot \dot{\gamma}^n \qquad (7.1)$$

the coefficients can be determined and the critical shear stress for the yield stress calculated (corresponding to Fig. 7.8: $C1 = \tau_0$; $C2 = K$; $C3 = n$).

In the tangent method, the abscissa and ordinate are interchanged in the plot (Fig. 7.9). It can be readily seen that the curve shows two linear regions of different

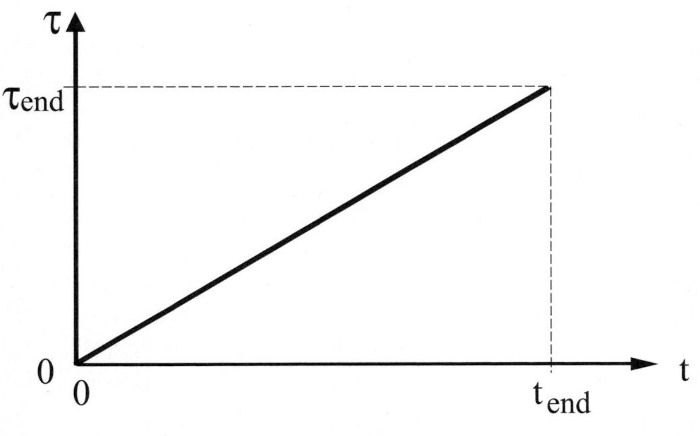

Fig. 7.7. Determination of the apparent yield stress

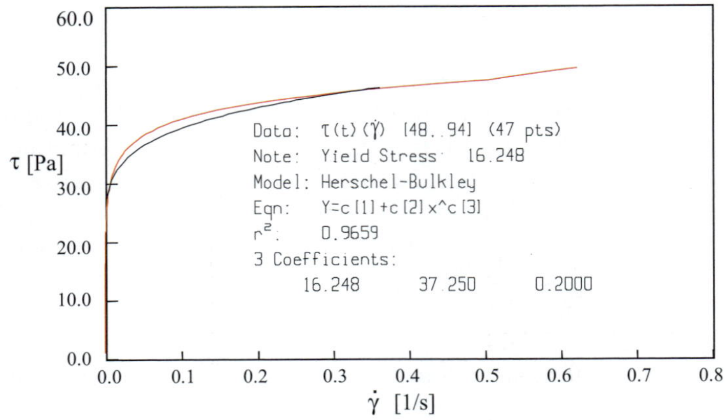

Fig. 7.8. Interpretation of the yield stress according to Hershel Bulkley

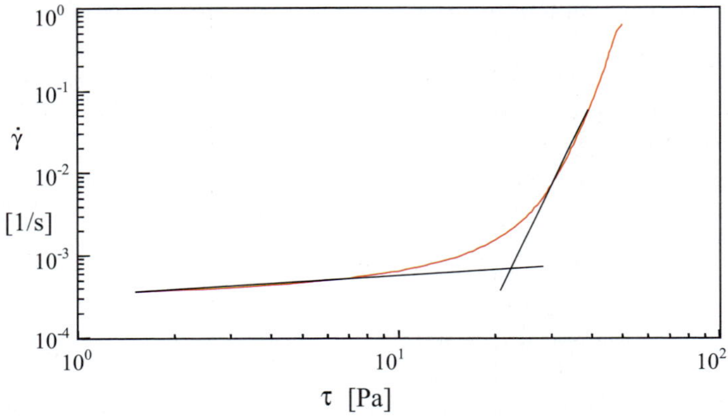

Fig. 7.9. Interpretation of the yield stress according to the tangential method

slope and a transition region. The intersection of the two tangents to the apparent linear regions is interpreted as the yield stress. This method is strongly dependent on the choice of measuring points for each tangent.

A third method for plotting the results [63] of a stress ramp test is the double logarithmic plot of the viscosity vs the shear stress.

In practice, plotting the viscosity as a function of shear stress gives good results. The viscosity initially increases to a maximum (Fig. 7.10) and then decreases again. The stress corresponding to the viscosity maximum is called the critical shear stress $\tau_{critical}$. This is the stress needed to cause the system to flow. Note that the calculated viscosity is an integral of the force over time. Therefore it is the shear stress rather than the viscosity that is the more important parameter for determining the apparent yield stress. In this way, discrete critical shear stresses can be assigned to specific products.

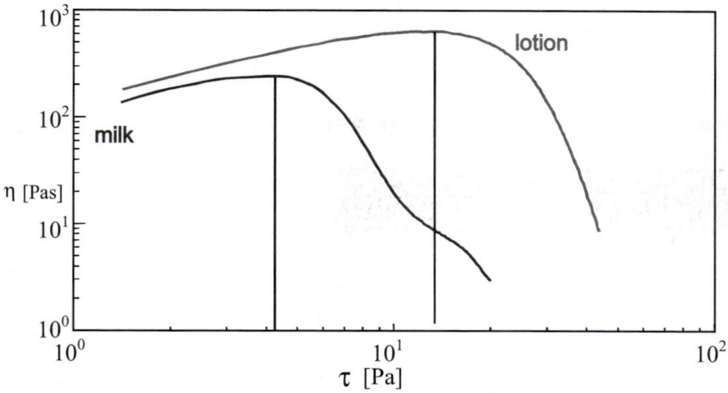

Fig. 7.10. Apparent shear stress of two products

7.2 Newtonian Flow Behavior

Products containing surfactants usually exhibit Newtonian flow behavior (Fig. 7.11). This is typically detected in the lower shear rate range. If flow behavior is Newtonian, the viscosity at constant temperature is independent of the applied stress or the velocity gradient. The value obtained is the mean of the viscosities over the velocity gradient.

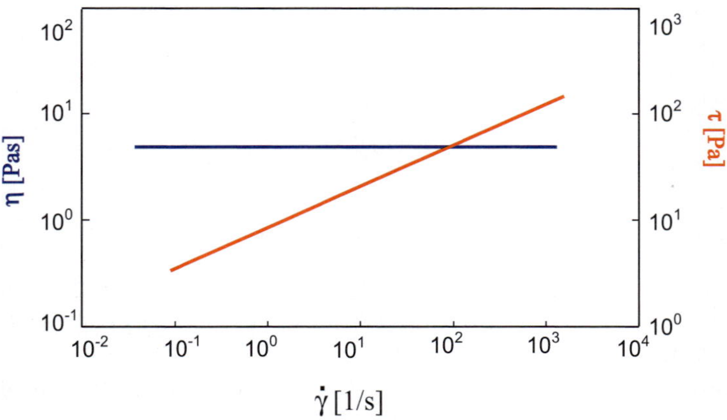

Fig. 7.11. Newtonian flow of a surfactant-containing product

7.3 Creep Test and Creep Recovery

The creep test is a simple and quick test for obtaining initial information on the viscoelastic properties of a sample from viscosity-relevant (as opposed to oscillation measurements) data. In this experiment (Fig. 7.12) a constant force

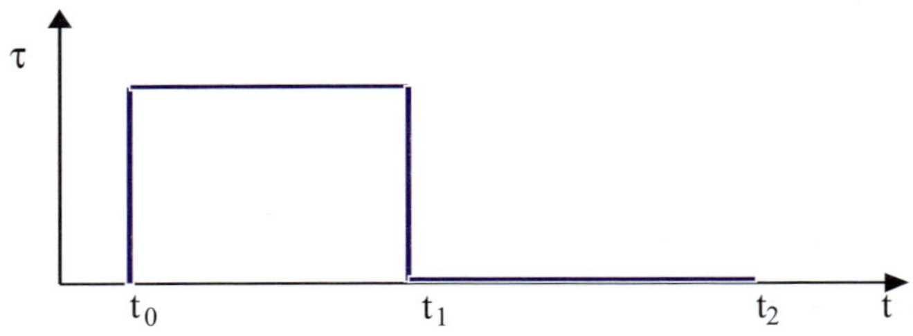

Fig. 7.12. Creep and creep recovery test

(shear stress) is applied to the sample at time t_0 and removed again at time t_1. The recovery up to time t_2 is recorded.

The sample responds initially to the force applied at t_0 with deformation. In other words, it starts to creep. At t_1 (after removal of the force), the sample recovers again. There are three different types of creep and creep recovery curves.

7.4 The Ideal Elastic Behavior

The first case we will consider is an ideal elastic body as exemplified by a steel spring. If a force is applied to the spring it responds with a deformation but returns to its original state after the force is removed (Fig. 7.13).

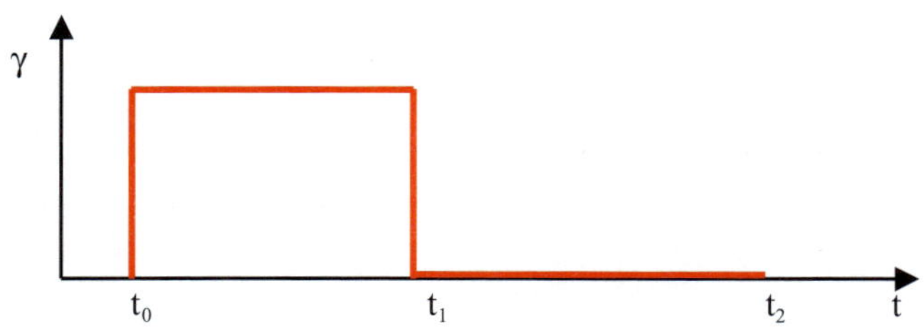

Fig. 7.13. Creep and creep recovery of an ideal elastic body

If the force τ is doubled, the deformation γ also doubles. Ideally, the energy stored in a spring will be recovered 100%. A body with these properties is also known as a Hookean body.

7.5 The Ideal Viscous Behavior

The second case we will consider is water as an example of an ideal fluid. The force τ applied to the fluid causes a linear deformation γ over time. In other words, the sample begins to flow.

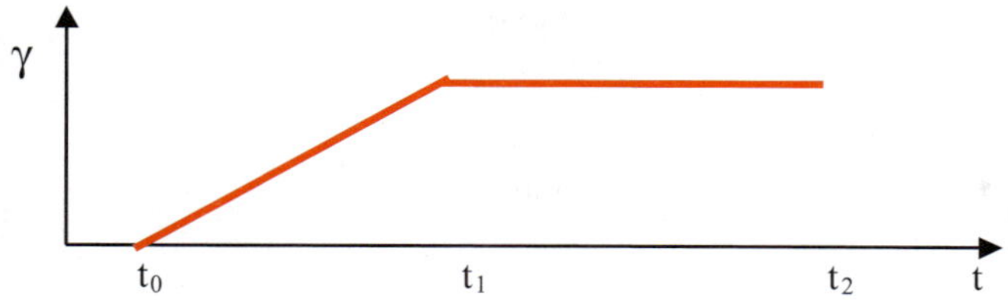

Fig. 7.14. Creep and creep recovery of an ideal vicous body

If the force is removed from the sample (Fig. 7.14), the deformation attained at this time (in our example t_1) is fully retained. The model in this case is the dashpot model according to Newton.

7.6 Real Viscoelastic Behavior

A real body is both viscous and elastic. This means that when a force is applied at the time t_0 deformation begins to take place much more slowly and, if we wait long enough (until t_1), the curve will approach a constant slope.

When the force is removed, part of the energy stored in the body will be released. The result is a recovery of the elastic part γ_e and a permanent deformation (Fig. 7.15) of the viscous part γ_v. A viscoelastic solid will therefore recover after a time lag but it will do so almost completely.

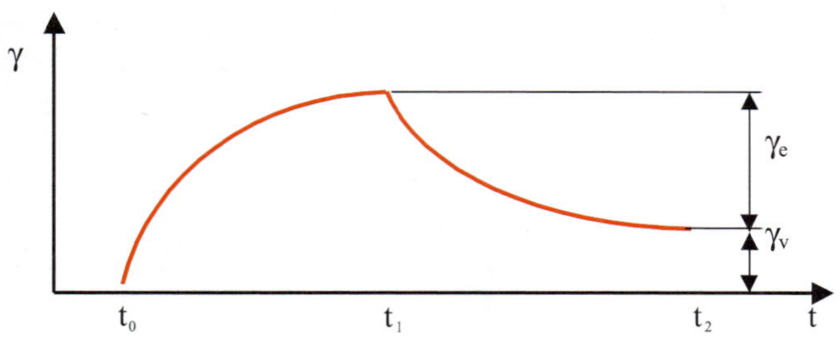

Fig. 7.15. Creep and creep recovery of a real viscoelastic body

7.7 Steady Flow Curve

The best way to measure the viscosity of a non-Newtonian sample at constant temperature and known shear rate is to progam a time test. Since both the measuring instrument and the sample need a finite time to reach constant conditions, in other words until the whole system is in equilibrium (Fig. 7.16), it is necessary to wait a certain time before measured values are obtained that can be evaluated.

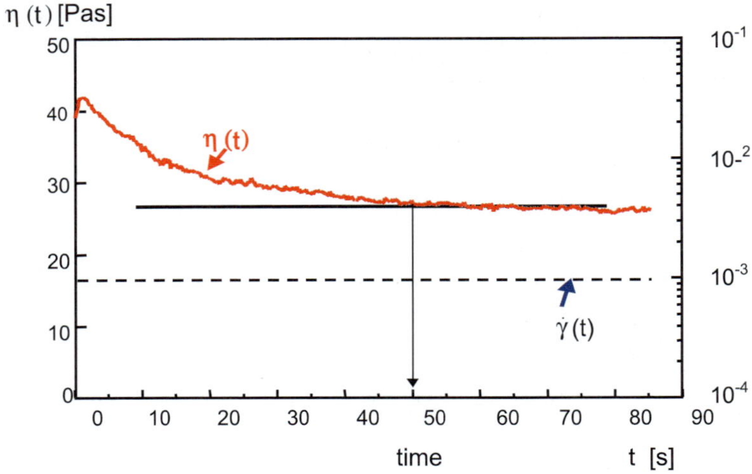

Fig. 7.16. Approach to equilibrium in static measurements

In the extreme example shown in Fig. 7.16, the viscosity increases within the first few seconds at 25 °C and a shear rate of $0.001\,s^{-1}$. After passing through a maximum, steady state conditions are not reached until after about 50 s and the viscosity measurement can begin. If several shear rates are applied sequentially in a step test (Fig. 7.2), the following results are obtained.

In the first segment of Fig. 7.17 at a shear rate of $\gamma = 0.01\,s^{-1}$ steady state conditions had still not been achieved after 150 s. At $\gamma = 0.1\,s^{-1}$ constant values could not be measured until after 75 s. At $\gamma = 1\,s^{-1}$ this was already the case after 25 s. Consequently, the greater the shear rate is the sooner steady state conditions

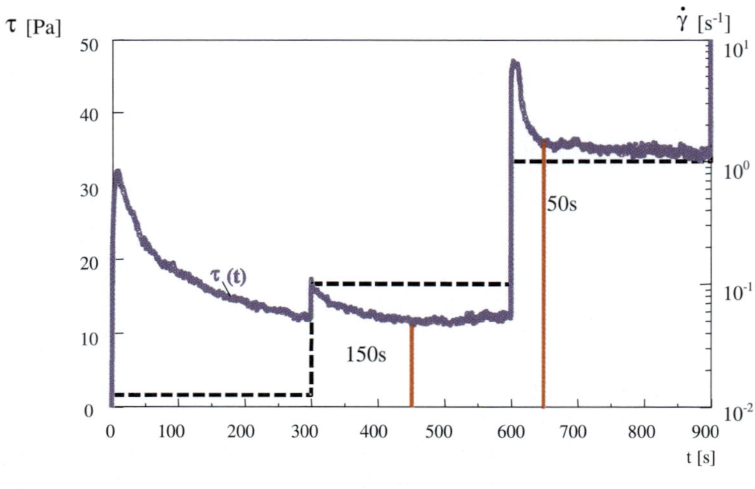

Fig. 7.17. Steady viscosity curve for non-Newtonian samples

will be reached. In other words, the lower the shear rate needed for a measurement the longer the measuring time will be and vice versa.

If we want to measure the static viscosity at higher temperatures with low shear rates, the relatively long measuring time is a considerable disadvantage because the sample can begin to dry out during the measurement (Fig. 7.18).

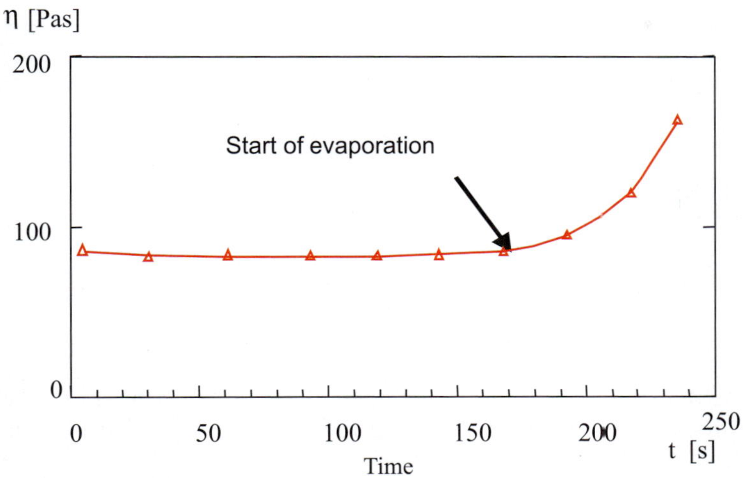

Fig. 7.18. Evaporation at higher temperatures

It is extremely important to keep as many boundary conditions as possible constant when measuring the viscosity because the viscosity depends on many factors including:

- Shear gradient
- Time
- Temperature
- Density
- Solids content
- Molecular weight

This makes it relatively difficult to determine why the viscosity changes during a measurement. Rheological measurements measure the effect and not the cause. Many measurements and the experience gained from them are needed to be able to interpret the data correctly.

7.8 Amplitude Dependence

Now let us consider oscillation measurements. This type of measurement provides information on the structural properties of materials. There are four different measuring variations:

- Amplitude variation
- Frequency variation
- Time dependence
- Temperature dependence

The amplitude variation, also known as the strain test, is performed at constant temperature and frequency (Fig. 7.19). Starting at small amplitudes, the strain is increased in small steps.

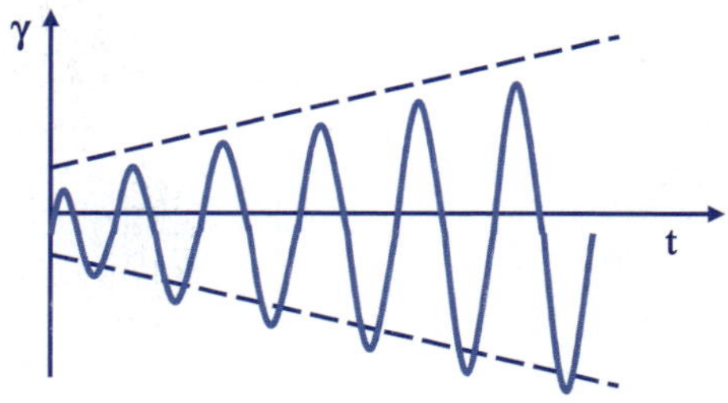

Fig. 7.19. The amplitude or strain test at T = const. and ω = const.

Obtained as the response are the two moduli G' and G'', which run nearly parallel in the lower frequency range. At a product-specific frequency, the response is no longer linear. The linear range is also called the linear viscoelastic range (LVR).

In the example shown in Fig. 7.20 the experiment was run at a frequency of $\omega = 100\,\text{rad/s}$ and $T = 25\,°\text{C}$. The plots of the moduli G' und G'' vs frequency are double logarithmic. At a strain of approximately $\gamma \approx 0.4\%$ (in this example 100% strain corresponds to 0.5 rad) it can be seen that the storage modulus G' leaves the linear range. Why do we need to perform the measurement in the linear viscoelastic range?

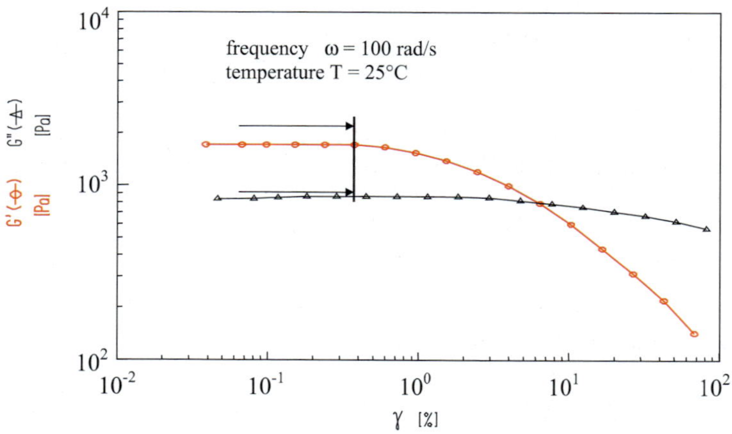

Fig. 7.20. Determination of the linear viscoelastic range with the amplitude test

There are two reasons:

1. Two measurements can be compared only if the same boundary conditions were maintained for both measurements. However, if the amplitude leaves the linear range, different results will be obtained in the analysis.
2. The second reason has to do with the measuring instrument. Even the most modern technology has its limits and in this case it is the torque resolution. A value measured near the torque resolution will be associated with an error of 100%. The minimum accuracy required for the measured value is $< 10\%$. This means the frequency range must be limited as appropriate.

The amplitude test (strain test) is performed as a kind of screening test for every new sample. As we will see later the results obtained for most samples to be measured are frequency dependent. Therefore the question of the correct test frequency arises. To be absolutely safe this test should be performed at the lowest and highest frequency, namely at the limits of the frequency test. This procedure must be repeated for every temperature used.

7.9 Structure Breakdown and Build Up

Most cosmetic emulsions are deformed when a small amount is removed. How quickly the structure is restored can be crucial. To determine this, the amplitude test must be slightly modified. First the linear viscoelastic [64] range described above must be determined. The test is run for a defined period at the amplitude determined in this test (Fig. 7.21) with T and ω constant.

Then the amplitude is suddenly increased for example 100-fold and the test continued at this deformation for about 5 min. Afterwards the amplitude is returned to the starting amplitude determined in the screening test and the sample observed for another 30 min.

We will again look at the storage modulus G' and loss modulus G'' plotted as a function of time as the response parameters (Fig. 7.22). In the first segment the storage modulus G' is larger than G'' (linear viscoelastic range). After the sudden increase in amplitude it can be seen that G' and G'' respond by decreasing dramatically (non-linear range).

It is striking that the relationship is now reversed and G' is smaller than G''. That would mean the sample flows. To get ketchup out of a bottle you have to

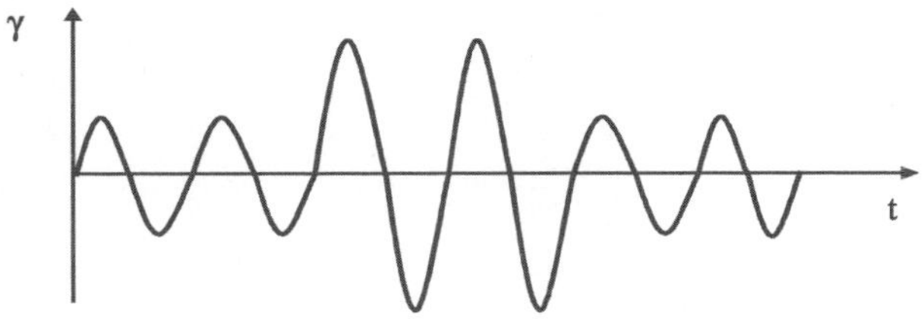

Fig. 7.21. Amplitude test for determination of structural breakdown and regeneration

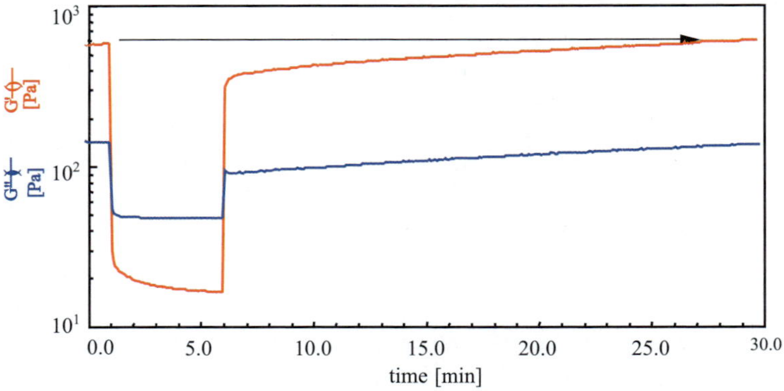

Fig. 7.22. Structural breakdown and build up for ketchup

shake it vigorously to make it flow. This is in fact exactly what is happening in this example, for the measurement shows us the rheological behavior of ketchup when it is shaken. If the bottle is then left to stand long enough the ketchup again thickens and will have to be shaken again to make it flow. This is what we see in the plot. After we return the amplitude to a value in the viscoelastic range it takes approximately 20 min for the original state to be achieved.

7.10 Time Dependence

Using dynamic-mechanical time-dependence measurements structural changes can be detected without superposition of shear (as in the static time test). The sample is observed for a certain time at constant temperature, frequency and amplitude (once again the LVR value) (Fig. 7.23).

In Fig. 7.24 the test result is presented as a plot of G' vs time. After approximately 400 s the absolute value of the storage modulus slowly increases. This can have several causes. For instance, the system could cross-link or some of the sample could evaporate, which is more likely for an emulsion.

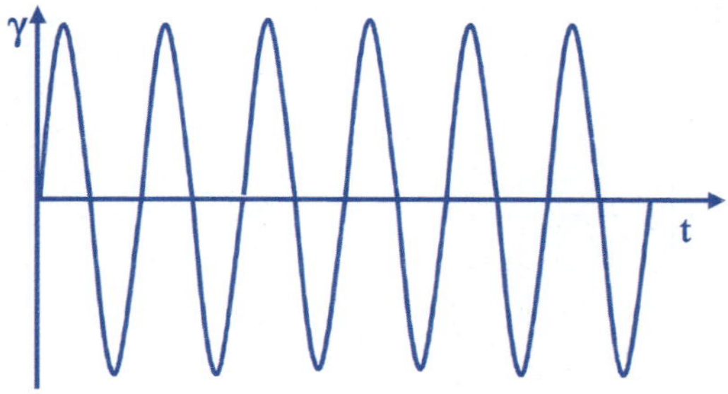

Fig. 7.23. Time dependence with T, ω, γ = const.

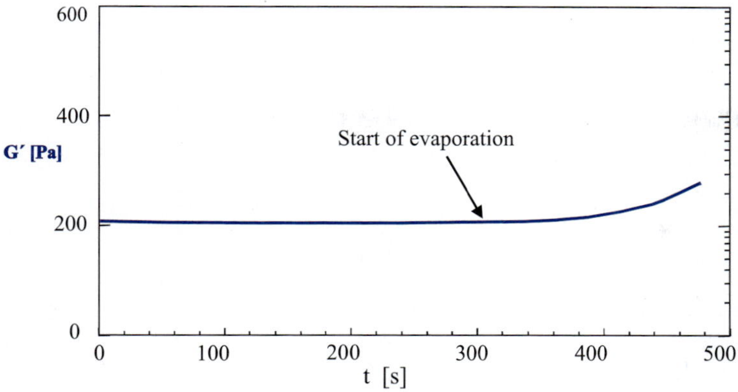

Fig. 7.24. Time dependence of the storage modulus with T, ω, γ = const.

7.11 Frequency Test

After the max. amplitude has been determined from the LVR, the frequency dependence is studied. Testing usually begins at the highest frequency and is then reduced in logarithmic steps. Starting at high frequencies has the advantage that the first measured values can be obtained very quickly because once again the following applies:

Low frequencies = long measuring time
High frequencies = short measuring time

In Fig. 7.25 the sinusoidal curve of the frequency test is presented starting at low frequencies. Obtained as the response in the double logarithmic plot are the two moduli G' (storage modulus) and G'' (loss modulus) or tan δ as a function of frequency (Fig. 7.26). Alternatively, the complex dynamic viscosity can be plotted vs frequency.

This example clearly shows that the complex dynamic viscosity η^* (*black, open circles*) decreases linearly with increasing frequency in the measured frequency range as is known from static measurements. The storage modulus G' (*blue triangles*) is always larger than the loss modulus G'' (*red, open squares*) in the measuring window. This shows that the sample must be present in the semi-solid state because

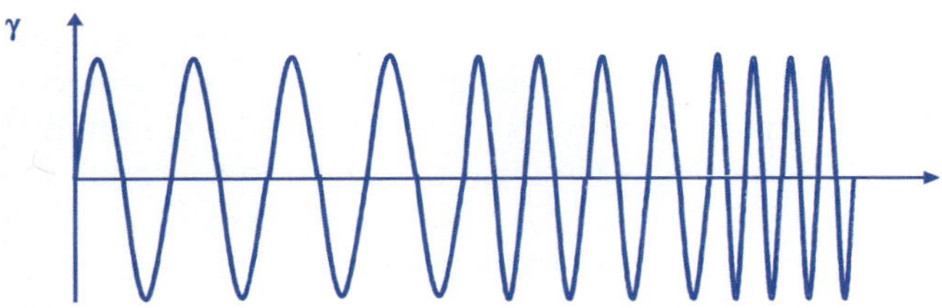

Fig. 7.25. The frequency test at constant temperature and amplitude

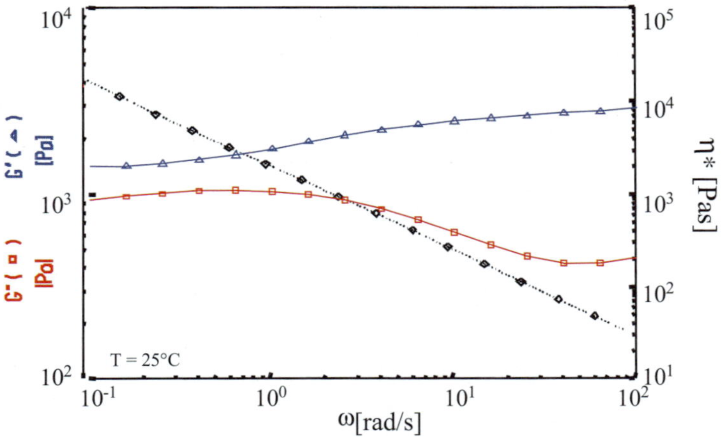

Fig. 7.26. Storage/loss modulus and complex dynamic viscosity as a function of frequency

the loss modulus G'' would always be larger than the storage modulus G' if it were a fluid.

7.12 Temperature Dependence

Often cosmetic emulsions are exposed to variable temperatures. For example, in summer a hand cream in the glove compartment of a car can easily reach a temperature of $T = 50\,°C$ during the day and drop to room temperature in the garage at night. In winter the temperature the opposite is true, with temperatures dropping to as low as $-20\,°C$. The hand cream is exposed to this stress with no other mechanical forces applied. Therefore, the method of choice is once again a dynamic-mechanical measurement. In order to measure the influence of temperature [65], first the LVR needs to be determined at the final temperature and attempts made to find a single value valid for the entire range.

If this is impossible the experiment must be divided into several parts. The boundary conditions are very simple: all parameters except the temperature are held constant (ω, γ = const.). We will perform two experiments. Both will start at $T = 25\,°C$. In one case the temperature will be lowered (Fig. 7.27) and in the other increased (Fig. 7.28).

In Fig. 7.27 it can be clearly seen that in the cooling experiment both the storage modulus G' and the loss modulus G'' suddenly increase at a temperature $T = -8\,°C$. This is typical of an oil-in-water emulsion because the external phase – water – freezes. In the heating experiment (Fig. 7.28) both moduli decrease at a temperature $T = 50\,°C$. This is where the internal structures of the emulsion sample begin to soften. Using these two experiments the temperature range can be determined in which the emulsion will not change during the measurement.

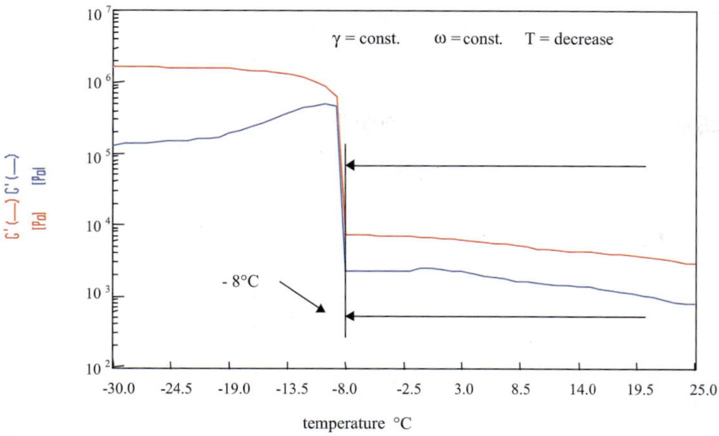

Fig. 7.27. Freezing point determination by temperature reduction

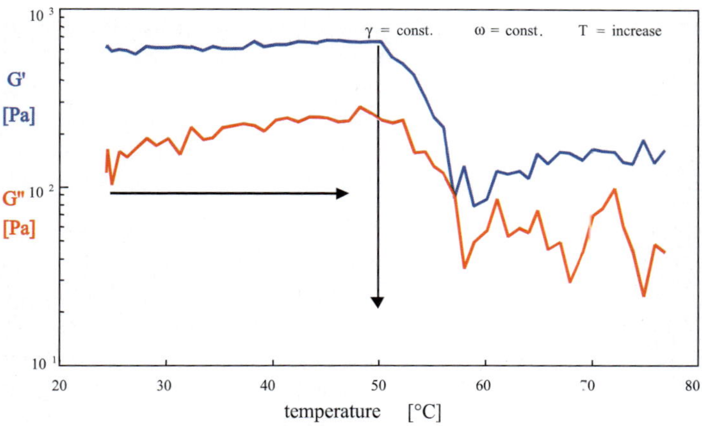

Fig. 7.28. Determination of the softening point by temperature elevation

7.13 Combined Temperature-Time Test

If these two tests are repeated at longer and shorter intervals and the absolute values of the storage and loss moduli compared over time, no differences should be detectable for stable emulsions. This is of course a time-consuming procedure. We must also consider whether the moduli G' and G'' will change if the cooling and heating curves are subsequently measured backwards, i.e. when heated or cooled, respectively. The test is very simple [66]: simultaneous negative and positive temperature ramps (Fig. 7.29) are programmed at constant frequency ω and constant amplitude γ. For example, the temperature can be increased to 50 °C from 25 °C.

Afterwards the sample is cooled to −10 °C with a constant cooling rate and subsequently heated again to 25 °C. These ramps are run several times. Once

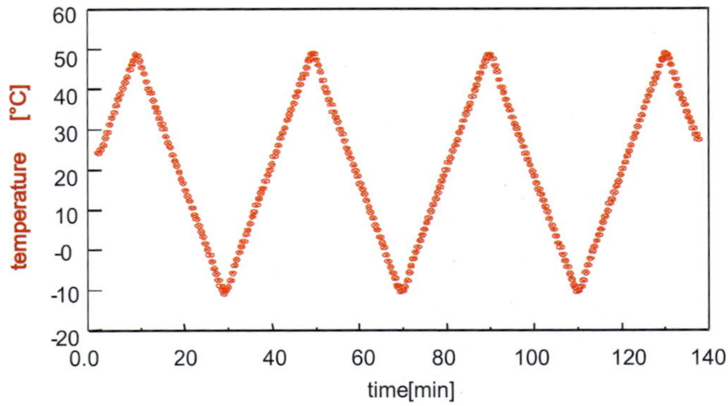

Fig. 7.29. Combined temperature-time test; also known as the cycle test

again, the limiting temperatures of the LVR must be determined. The combined temperature-time test is also known as the cycle test.

In Fig. 7.30 we see the result of a cycle test. The storage modulus G' is always larger than the loss modulus G'' over the entire measuring window. If the storage modulus G' increases so does the loss modulus G'' (but somewhat more). The whole procedure is repeated as often as the temperature ramps are run. In the cold the moduli increase slightly as they approach the freezing point. At higher temperatures a slight softening of the structure begins to become apparent.

This can be seen even more clearly in the plot of the quotient $G''/G' = \tan\delta$ vs the temperature (Fig. 7.31). It can easily be seen that the temperature cycles do not affect the sample. This sample can be said to be temperature resistant for the temperature range from $T = -10$ to $+50\,°\mathrm{C}$.

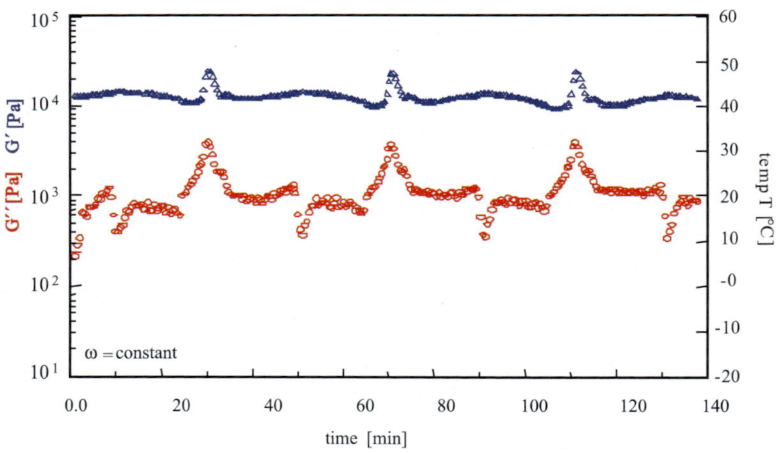

Fig. 7.30. Storage and lost moduli G' and G'' in a cycle test

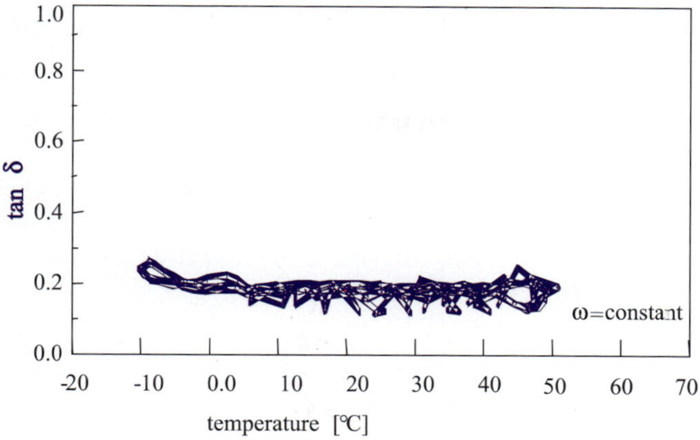

Fig. 7.31. The cycle test in the tan δ vs temperature plot

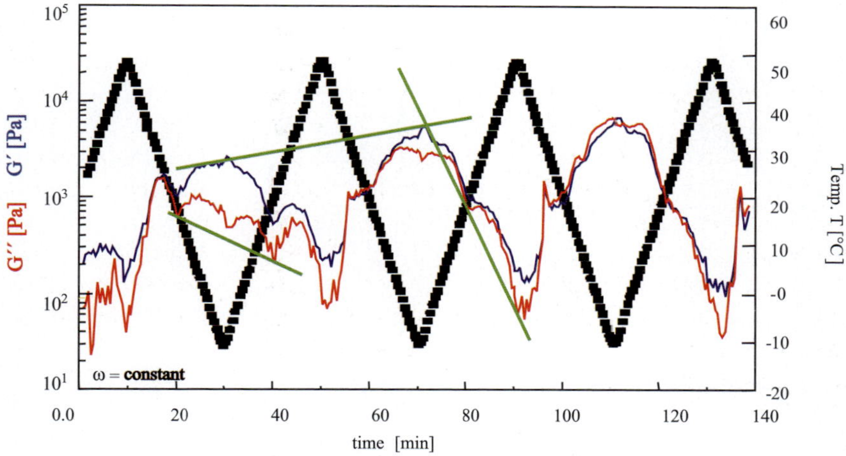

Fig. 7.32. The cycle test for an unstable formulation

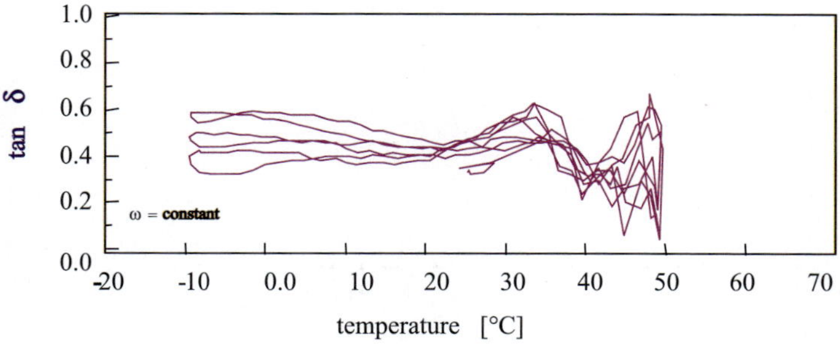

Fig. 7.33. The cycle test for an unstable formulation

In Fig. 7.32 we see the result of another cycle test. In this case there is a change in the absolute value of the storage modulus G' and the loss modulus G'' at corresponding temperature changes over the four cycles.

At which temperature the mean problem is can be seen more clearly in the $\tan\delta$ plot vs. the temperature (Fig. 7.33).

At higher temperature the curve is not so clear, but there is no change to be seen over the cycles. In the cold range between $-10\,°C < T < +10\,°C$ you can see hysteresis. That means the sample will have problems in the cold and therefore we say that this is the behavior for a temperature unstable sample.

8 Analysis of Measuring Results and Correlations with Other Tests

In this chapter we will discuss the results of rheological measurements. An experiment can be plotted in several different ways, as described in the preceding chapter, and analyzed accordingly. This can sometimes result in dramatic differences. If mathematical models are used, the analysis may be very confusing to the non-expert. Therefore it is impossible to give a general answer to the question of how a certain experiment should be analyzed. Every user must decide this individually and then always state the defined boundary conditions. Here the possible correlations of measured data with user-specific, and process- and user-relevant tests will be presented.

8.1 Yield Stress

An unknown sample will usually first be subjected to a stress-time ramp test. The result of this test is called the onset of flow, yield point or yield stress, as described in Sect. 7.1. Several things need to be considered. Why is it good to know the maximum viscosity or the associated critical stress? If a set shear stress, meaning a force, is applied, the distance traveled is obtained as the response, as we learned in Chap. 5. What good is this to the consumer? To answer these questions we will perform a theoretical experiment with two different samples. They will be two products, a cream and a body lotion. Most creams are stored in jars or cans. Let us take the lid off the cream can and remove the protective seal. If we then turn the can upside down, what happens? The answer is: "nothing!" The structure or structural strength of a cream is so great that it stays in the can. Now let us fill another can with the body lotion and repeat the experiment. This time nearly all of the contents of the can ends up spread all over the floor.

Let us now do a second theoretical experiment with the two samples. We want to remove the cream or body lotion from the can. Using our universal rheometer, our hand, we discover that the force we need to remove the cream is markedly higher than that needed to remove the body lotion. This is also exactly what the consumer expects. A cream should clearly be thicker than a body lotion. This also means that the consumer associates the terms cream, lotion and milk with certain properties. It is almost impossible to sell the consumer a cream as a lotion and likewise impossible to sell a lotion as a cream.

Let us now return to the analysis of shear-time ramp tests and possible correlations with other property relationships. Both theoretical experiments show that that a very specific force (product dependent) is needed to cause a sample to flow.

8.1.1 Correlations of the Yield Stress with the Primary Skin Feel

Now we want to make use of the experience of trained panelists who can evaluate the skin feel of different products. The sensations experienced by these panelists can be summarized as follows.

Whether a cream will have a good skin feel depends on several rheological factors. A high yield stress is desirable when removing a cream from the jar and at the start of application, but this should be quickly exceeded during the course of further application. At the end of application when the cream has been rubbed on the skin a low viscosity is perceived as optimal because the cream should be absorbed quickly. Let us now take a closer look at the period of initial application also called the primary skin feel.

The term primary skin feel [67,68] comprises the sensations perceived near the yield stress and at the start of distribution on the skin. To study this more closely, several lotions (samples A to G) and two creams (samples K and L) were subjected to a panel test of this type and the respective rheological data measured as well. It is important to establish rules for conducting a panel test. A defined number of persons must apply a defined volume of the samples to a specific skin area such as the forearm. There are now two evaluation options: a verbal description allows a totally free evaluation of the sample by the panelist. The disadvantage is that it is relatively difficult to summarize the results of several panelists. The second option is to rank the products according to defined criteria. Panel tests are very time-consuming because a skin site should not be used more than twice a day for the test. Otherwise the skin will become very stressed, which may influence the condition of the untreated skin.

To correlate the primary skin feel with rheological material properties, the sensory evaluations were compared with the yield stress and viscosity maximum obtained from a shear stress time ramp test. The shear stress was increased logarithmically with time to expand the region of the yield stress and yet be able to utilize and present the entire measuring range of the instrument. In this experiment the change in the shear stress and shear gradient occurring in the sample were unknown. Because the results of this dynamic test procedure depend on how quickly the shear stress is increased, an optimal rate of increase must be determined that allows good discrimination of the yield stress of similar products.

To determine when a fluid begins to flow with a high sensitivity, the steady flow curve can be studied because the yield stress is already exceeded even at the smallest shear gradients. In contrast, if a stress ramp is used to determine the onset of flow, the yield stress τ_F can be read at the maximum η max of the viscosity curve.

A plate/plate system with a diameter of $d = 25\,\text{mm}$ and a gap of $h = 1\,\text{mm}$ was used for the measurement. The instrument was programmed so that the shear stress increased from 0 to 400 Pa in 2 min. Optimization of this program will be

Table 8.1. Comparison of the yield stress and viscosity maximum of lotions A–G and creams K–L with the sensory evaluations of the skin feel

Product	Yield stress τ_F [Pa]	Viscosity maximum η [Pas]	Sensory evaluation (1: very good ... 5: poor)
Lotion A	13	500	3.0
Lotion B	12	510	3.0
Lotion C	12	460	2.5
Lotion D	0.5	250	2.5
Lotion E	10	300	3.0
Lotion F	12	570	2.7
Lotion G	6.5	120	3.0
Cream K	15	3500	2.5
Cream L	8.5	1100	2.0

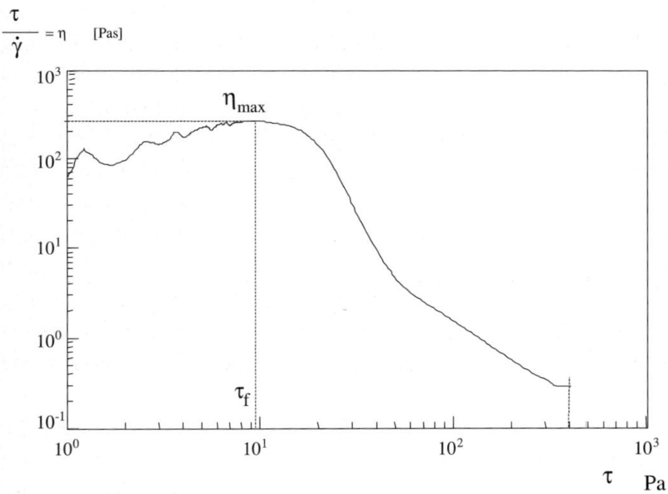

Fig. 8.1. Dynamic viscosity curve for Lotion D for study of the yield stress

discussed in the next section. The measured results were obtained in tabular form and as a diagram. The data from the table were used for the analysis and showed the yield stress for Lotion D at $\tau_F = 9.5\,\text{Pa}$, as can be seen in Fig. 8.1. As a dynamic measured value, the yield stress depends on the slope of the chosen shear stress ramp.

8.1.2 Optimization of the Stress Ramp Test

To optimize the slope of the shear stress ramp for measurements of the yield stress, Lotions D and E were measured. Each product was studied with preset shear stress ramps ranging from 12.5 Pa/min to 400 Pa/min. In Fig. 8.2 the dynamic viscosity curve is plotted as a function of the shear stress for Lotions D and C for different shear stress ramps, with the region of the yield stress expanded.

Shown in Fig. 8.3 is the same test for the Creams K and L, where the maximum number of rotations per minute of the rheometer limits the maximum shear stress to 200 Pa. The test results in Figs. 8.2 and 8.3 confirm that the viscosity maximum and yield stress depend on the test procedure.

To measure the dynamic viscosity curve as quickly as possible and improve discrimination, a shear stress ramp of 200 Pa/min can be used for the samples. Shorter measuring times reduce discrimination and longer measuring times provide no further improvement. All other measurements of the yield stress – as well as those described in the preceding section – were performed with a shear stress ramp of 200 Pa/min.

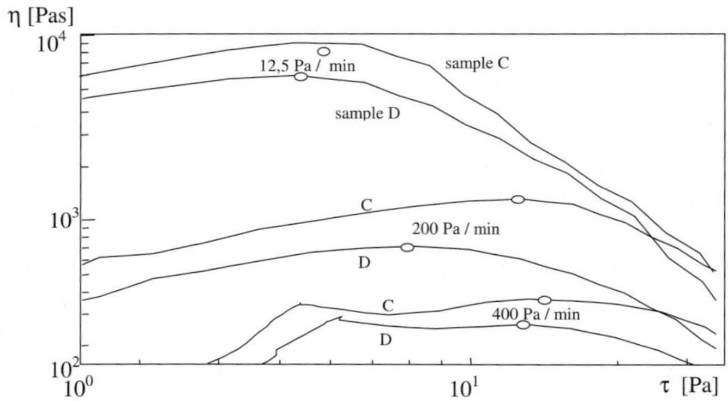

Fig. 8.2. Expanded plot of the dynamic viscosity curves for Samples C and D and their maximum as a function of shear stress for different shear stress ramps

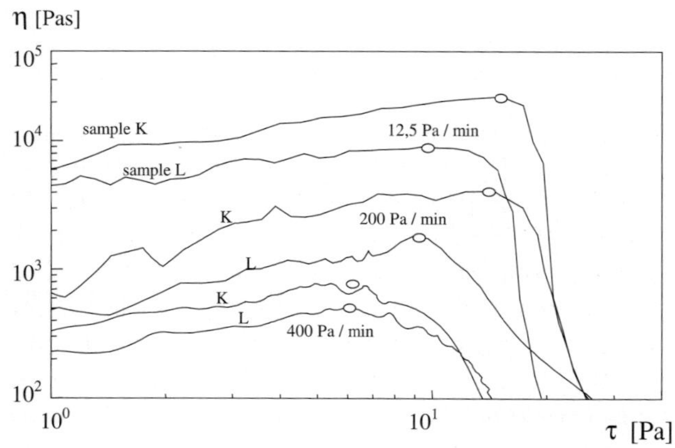

Fig. 8.3. Expanded plot of the dynamic viscosity curves for Samples K and L and their maximum as a function of shear stress for different shear stress ramps

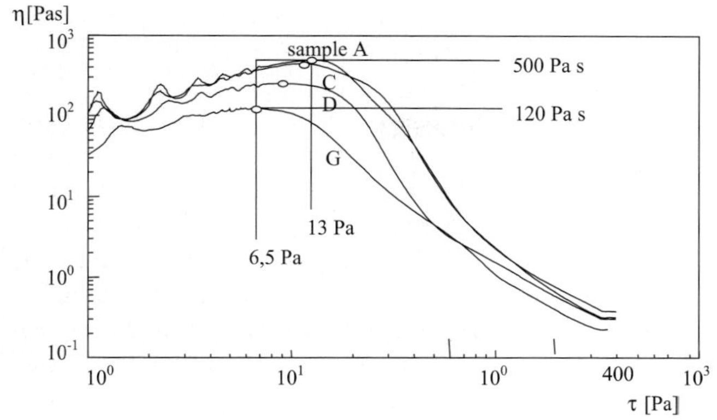

Fig. 8.4. Dynamic viscosity curves for the Lotions C and D assessed as good and for Lotions A and G assessed as poor

If we now compare the sensory data with the yield stress and viscosity maximum data and plot the curves measured for the samples assessed as good as well as those with the lowest yield stress (Fig. 8.4) a window of sorts becomes apparent. Limits can be defined within which the viscosity maximum must be found for the sample to be assessed to have a good primary skin feel.

Lotions C and D, which received the best assessments, conform to the range limits. Lotion E conforms to the limits for the yield stress and viscosity maximum but not to the limits for the viscosity minimum. Lotions A, B, F and G do not conform to at least one of the range limits.

In this example, for a lotion to be assessed with a good primary skin feel its viscosity must be in the range $120\,\text{Pas} < \eta < 500\,\text{Pas}$ and its yield stress should be in the range $6.5\,\text{Pa} < \tau < 13\,\text{Pa}$. A lotion that is supposed to have an optimal skin feel assessment must conform to all limits. These limits must of course be determined for another product class such as creams according to the same procedure.

8.1.3 Residue Emptying

Some products are filled into pump dispenser bottles because of their low viscosity. In this case the task is to find out whether the residue emptying behavior [69] can be determined by means of rheological measurements. This is especially important to consumers because they have the right to know that they can use all of the contents of the bottle for which they paid. To be able to guarantee optimal emptying of residues, the formulator must make sure that the product to be pumped always flows to the bottom of the bottle and does not adhere to the wall because the end of the dip tube through which the product is pumped is located near the bottom.

The left side of Fig. 8.5 shows the optimal behavior for a fluid to be pumped. It always flows to the bottom. These are usually fluids with Newtonian flow behavior. But what about real cosmetic emulsions that sometimes have a yield point? To answer this question, five products were studied in a pumping test.

Fig. 8.5. Schematic representation of a sample with good and poor residue emptying

Table 8.2. Comparison of the residue emptying behavior of five products with their viscosities at $T = 25\,°\mathrm{C}$ and a shear rate of $\dot{\gamma} = 10\,\mathrm{s}^{-1}$

Sample	Residue emptying	Viscosity [Pas] at 25 °C and $10\,\mathrm{s}^{-1}$
Product A	Poor	7000
Product B	Good	6000
Product C	Good	2000
Product D	Moderate	1500
Product E	Good	250

It is evident from the table that products B, C and E exhibit good residue emptying behavior. There appears to be a correlation with the shear rate $\dot{\gamma} = 10\,\mathrm{s}^{-1}$ at $T = 25\,°\mathrm{C}$ because Product A, which is the product with the highest viscosity ($\eta = 7000\,\mathrm{Pas}$), has the worst residue emptying behavior and the ranking shows – with the exception of Product D (moderately good residue emptying) – that the residue emptying behavior is good to moderate below a critical viscosity. Consideration of the entire flow curve as presented in Fig. 8.6, however, leads to another conclusion.

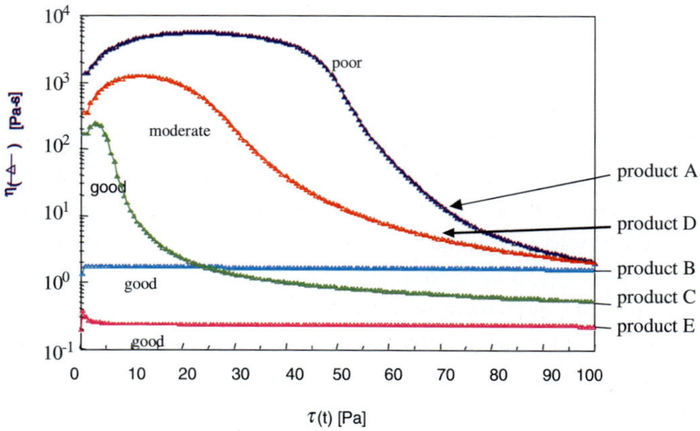

Fig. 8.6. Correlation of the yield stress with the residue emptying behavior of products A–E

Products B and E exhibit Newtonian flow properties and therefore show good residue emptying behavior, as would be expected. The three other products exhibit non-linear flow behavior with a yield point. Ranking according to the yield stress gives a logical result. When the yield stress of a product is higher than a critical shear stress τ_{crit} it can no longer be assured that the product will be completely emptied from the bottle. This can be thought of as follows: a high yield stress means that the structural strength of the product is also high, which causes the product to adhere to the wall of the bottle. The force needed to cause the product to flow is so large that the product does not begin to flow under its own weight.

8.1.4 Energy Input

At first glance preparation of an emulsion appears to be very simple. Two immiscible fluids are dispersed for instance by stirring vigorously and gives a (short-lived) emulsion. Depending on the composition of the two fluids either a w/o or an o/w emulsion is formed. However, because the new system has a higher free energy [70], it will separate again after a short while with release of energy. To stabilize such systems for longer periods emulsifiers are added that can delay separation into the thermodynamically more stable original state until the emulsion found its intended use.

The emulsion production process can be divided into three steps:

1. Pre-emulsification
2. Fine emulsification
3. Stabilization

In the pre-emulsification step the water and oil phases are mixed with heating and stirring to form a raw emulsion (premix) with large droplets. These are subsequently deformed by external shear forces during the fine emulsification step and are reduced in size when a critical deformation is exceeded. Emulsifiers at the phase interface stabilize the newly generated smaller droplets and prevent them from coalescing.

The emulsification process consists of the breakup of droplets and coating of the phase interface. Immediately after breakup the phase interface is no longer completely covered by emulsifier molecules. Adsorption of more surfactant molecules takes time and is dependent on the interface coating kinetics of the emulsifier system used. The coating density influences both the interfacial tension and hence the energy needed for further size reduction as well as the stability of the drops formed [24].

Inadequately stabilized droplets can coalesce with other droplets when they collide if enough time is available. The continuous phase between the colliding droplets must be displaced to a critical film thickness (film drainage). Coalescence can be prevented if the repelling forces between droplets are strong enough. These repelling forces are induced by adsorbed emulsifier molecules. Spreading of emulsifier molecules distributed unevenly over the droplet surface (Gibbs–Marangoni effect) slows film drainage and stabilizes the droplets even if the interface is still not completely covered [26].

Both droplet size reduction and coalescence of size reduced but still not completely stabilized droplets determine the emulsification result and the dispersity of the prepared emulsion.

8.1.4.1 Measurement of the Energy Input

Studies by Pedrocchi and Karbstein [24] on different disperser units have shown that the results of size reduction depend mainly on the energy density and viscosity of the dispersed phase. The type of energy input has no influence on the effectiveness of droplet size reduction for emulsifier units that utilize turbulent flow to break up the droplets.

Since the energy loss [71] is a function of the speed of the rotor, a no-load curve must be constructed to calculate the energy loss. Figure 8.7 shows the dependence of current input on the speed n for the empty homogenizer. Using this characteristic curve the energy loss can be calculated during production of emulsions from the rotational speed.

Lashmar and Beesley [72] studied the influence of a number of production parameters on the rheological properties of o/w emulsions. An influence of the homogenization speed and duration as well as the phase composition and order in which the phases are added could be demonstrated. To exclude an influence of the production conditions on structure, the emulsions studied were prepared under constant conditions.

Since the weight of the emulsions to be homogenized varies during the production process (removal of samples, addition of fragrance), the energy density (energy per mass) was calculated. Table 8.3 presents the energy input and percentage lost energy for the ten complex w/o emulsions prepared.

Figure 8.8 shows the dependence of droplet size on the energy input for 10 complex w/o emulsions in the double logarithmic plot. It can be seen that the droplet size decreases exponentially as a function of the energy density E/m (energy

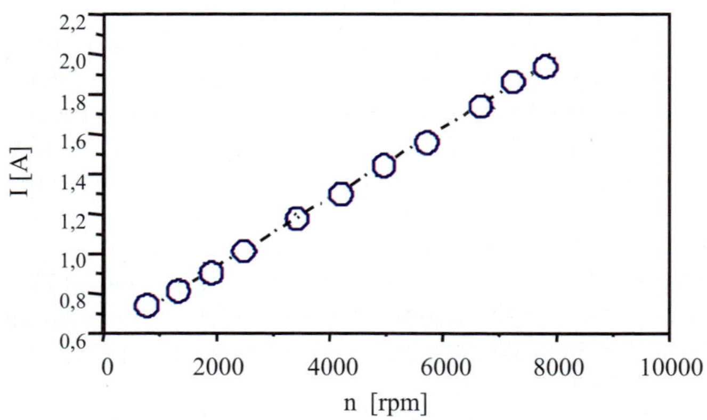

Fig. 8.7. No-load characteristic curve plotted as the current intake against the speed n

Table 8.3. Homogenization of w/o emulsions with different energy input

Emulsion	Energy density [kJ/kg]	$d_{v,50}$ [µm]	Yield stress τ [Pa]
A1	8.68	2.28	8.9
A2	14.34	1.28	14.2
A3	39.85	0.85	42.3
A4	87.77	0.33	91.4
A5	40.4	0.36	60.2
A6	67.95	0.35	65.4
A7	114.8	0.29	84.8
A8	22.87	0.86	28.2
A9	49.68	0.63	54.3
A10	102.12	0.31	87.6

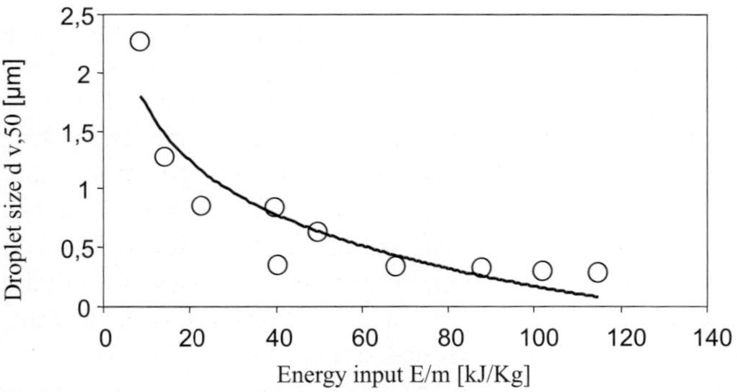

Fig. 8.8. Influence of the energy input on the mean droplet diameter of ten w/o emulsions

per mass). This dependence was described by Karbstein for different dispersing units and different dispersing zone geometries.

The influence of the energy input on the yield stress of complex w/o emulsions of constant formula composition is shown in Fig. 8.9. An increase in the input energy density by a factor of 15 increases the yield stress by a factor of 10. Accordingly, the internal structure increases with increasing energy input. This increase cannot be explained either by an increase in the phase volume ratio ϕ or by synergistic effects of ingredients that approach each other. Since the droplet size decreases with increasing energy input, the surface-volume ratio of the dispersed droplets is shifted to higher values. As a result, the number of emulsifier molecules adsorbed on the surface increases, resulting in an increase in interparticulate interactions for a constant phase volume ratio ϕ.

Therefore the yield point can be used to make integral statements on the internal structure of disperse multi-substance mixtures. However, they provide no

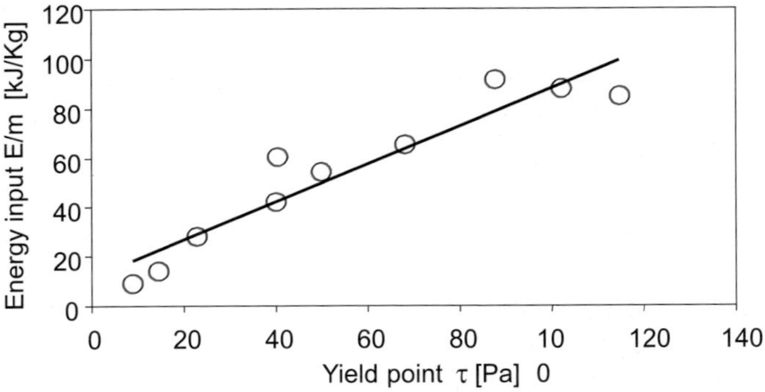

Fig. 8.9. Influence of the energy input on the yield stress τ_{yield} for ten w/o emulsions

information on the microstructure or the molecular causes of structural breakdown.

8.1.5 Droplet Sizes and their Distribution

To evaluate the influence of the energy input on the efficiency of droplet size reduction it is necessary to determine the droplet sizes and their distribution. This is done using static laser light scattering [73] (MasterSizer, Malvern Instruments GmbH). Frauenhofer diffraction has been described in detail in the literature [74, 75]. The measuring principle is based on the proportionality of the angle dependency of the scattered light on particle size. In measurements with the MasterSizer the diffraction spectrum of the particle population is recorded with the aid of 44 detectors. Use of an He–Ne laser (wavelength $\lambda = 633$ nm) allows measurement of particle sizes in the range from 50 nm to approximately 800 µm.

Measurement of particle sizes less than 10 µm are associated with large errors because the intensity of the scattered light can no longer by described by a simple equation [76]. Not only the size but also the shape, wavelength and optical properties of the particles influence the intensity of scattered light in this range. Consequently, the measured values always show a larger fraction of fine particles than is actually present in the sample. This measurement error is reduced in the MasterSizer in two ways: a wide angle detector (135°) also measures the scattered light from smaller particles. Using the refractive indices of the dispersed and continuous phases and the extinction coefficients of the disperse phase a correction factor (Mie correction) can be added.

Droplet sizes and their distribution can be described in a variety of ways. Depending on whether or not it is based on the number, surface area or volume, different values are obtained for the mean droplet diameter. To estimate the emulsifier requirement and for information on the specific surface area of the internal phase, the surface mean diameter (Sauter diameter $d_{1,2} = d_S$) is studied. For information on the creaming stability and coalescence behavior data on the distribution of the volume of the disperse phase are needed. The volume median diameter $d_{v,50}$ is more suitable for problems of this type [77].

To evaluate the efficiency of droplet size reduction the mean droplet size as well as the droplet size distribution should be known. A measure of the distribution span is obtained with Equation:

$$\text{span} = \frac{d_{v,90} - d_{v,10}}{d_{v,50}} . \tag{8.1}$$

Static laser light scattering offers several advantages over other methods as an absolute method for particle sizing. These include quick execution, a wide measuring range and no calibration. Since multiple scattering must be prevented, however, the samples must be highly diluted before measurement. The structural changes in emulsions caused by dilution are the biggest disadvantage of the method. In addition, it is difficult to measure refractive indices of interfaces at which emulsifier molecules are adsorbed.

To measure the droplet size, emulsions are diluted with isoparaffin. Used for the Mie correction is the refractive index of water and isoparaffin. To exclude the influence of dilution on changes in droplet size, the experiments are performed so that a sequence of four more measurements follows the first measurement. An influence of dilution on droplet morphology can be excluded if no change in the droplet diameter is detected with increasing time in solution.

In the double logarithmic plot, a linear relationship between the energy input and yield stress can be recognized. With increasing energy input the yield stress also increases.

The plot of the yield stress vs. the mean droplet diameter clearly shows that the shear stress needed to initiate flow of an emulsion increases with decreasing mean droplet diameter. It can also be seen from Fig. 8.11 that the mean droplet diameter determined with the yield stress method does not correlate with the results of laser scattering until a droplet size of $d < 2\,\mu m$.

Homogenization increases the energy density in an emulsion and results in a reduction of droplet size, which in turn results in a clear rise in the yield stress.

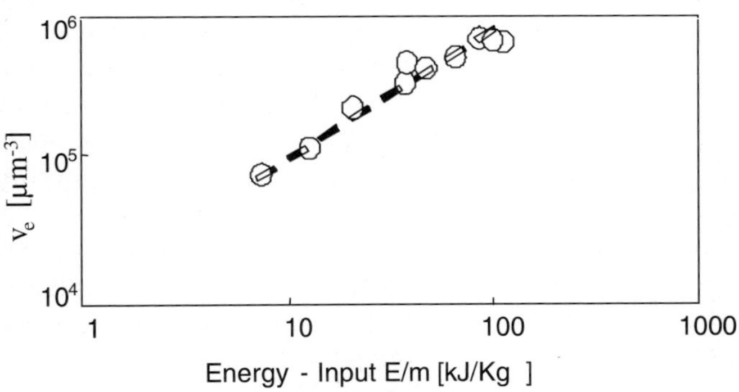

Fig. 8.10. Influence of the energy input on the volume median diameter of complex w/o emulsions

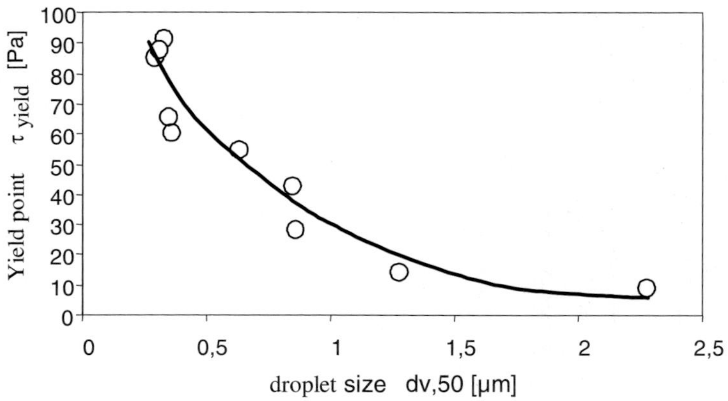

Fig. 8.11. Correlation of the yield stress with mean droplet diameter

8.1.6 Pumpability of Cosmetic Emulsions

Now we will consider how the viscosity of an emulsion can be used as a parameter for packaging optimization. Regardless of the product, every consumer wants to be able to use it all up. In the world of cosmetics there is of course a wide variety of product dosage forms [68], from a small jar or can, a tube or a hand pump to a flagon. Obviously, emptying a jar or can is fairly straightforward, but emptying a tube is another story. We will use an example to explain this. The mouth of a tube is nothing but a short capillary like that described in Sect. 5.6. The opening determines not only the product amount (larger diameter = more product) but also the force needed to squeeze the product through the opening. The greater the structural strength of an emulsion, the stronger the force needed to squeeze the product out through the neck of the tube.

The situation is much more complicated for a standard lotion pump. If we take apart the packaging we find, in addition to the outer casing, a long tube inside that extends almost to the bottom and a large plastic bead, a few small steel beads, a steel spring and a few other plastic parts in the a head – the actual pump. The lotion (in our example) is sucked through the dip tube (the tube) by applying pressure to the pump lever. Then the product is forced between the ball and a sealing edge through an annular gap. It takes several pump strokes to fill the entire system before any lotion sees the light of day. If we take a closer look at the pumping operations we see that the product is exposed to very different shear rates, relatively small ones when drawn into the dip tube and while being conveyed up to the annular gap, the highest shear rates occurring during the backstroke near the ball valve in the emerging gap (Fig. 8.12). Consequently we need to calculate first the shear rates of these processes in order to derive a shear gradient range from them. At the same time pumping experiments need to be performed with different products in order to correlate these results with each other.

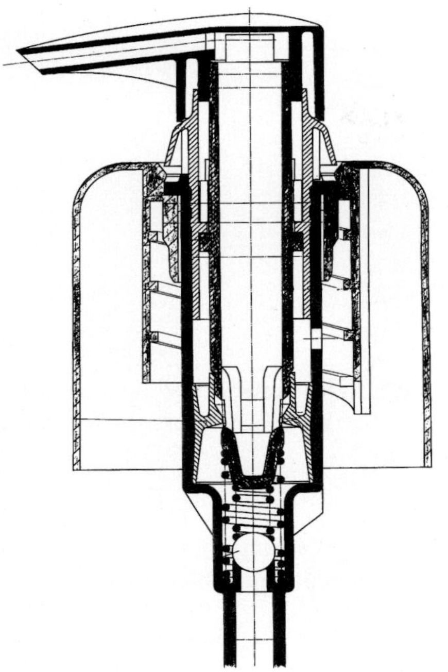

Fig. 8.12. Plan of a standard pump

8.1.6.1 Estimation of the Maximum Shear Rate

The apparent shear rate $\dot{\gamma}_{ap}$ in the dip tube can be described as follows according to Hagen–Poiseuille:

$$\dot{\gamma}_{ap} = \frac{4}{\pi \cdot R^3} \cdot \dot{V} \qquad \begin{array}{l} V = V/t = \text{Volume flow} \\ R = \text{Radius dip tube} \end{array} \tag{8.2}$$

The equation for calculation of the shear rates $\dot{\gamma}$ occurring in the annular gap between the ball and sealing edge is

$$\dot{\gamma}_{ap} = \frac{6}{\pi \cdot d' \cdot h^2} \cdot \dot{V} \qquad \begin{array}{l} d' = \text{Mean diameter of the annular gap} \\ h = \text{Ball stroke} \end{array} \tag{8.3}$$

Examination of the two equations shows that for the same volume flow (V/t) the shear rates will be larger in the annular ring than in the dip tube by at least a factor of 2 and at most a factor of 8.

The shear rates derived on the basis of these assumptions therefore range from 10^2 to $10^3\,\mathrm{s}^{-1}$ during the backstroke.

It is apparent from Table 8.4 that the correlation of the shear rate at $500\,\mathrm{s}^{-1}$ with the pumping behavior could show a trend. With decreasing viscosity the pumping behavior improves. Only the samples assessed with moderate and poor pumping behavior do not fit the trend. Before considering this approach any further we should first ask whether there might be another measuring variable that gives a better correlation. In fact, there is another correlation because we have here

Table 8.4. Experiments on the pumping behavior of cosmetic emulsions

Emulsions studied	Pumping behavior	Viscosity at 500 s^{-1}
Product A	Poor	1500 mPas
Product B	Good	250 mPas
Product C	Moderate	6000 mPas
Product D	Good	7000 mPas
Product E	Good	2000 mPas

primarily a mechanical variable, the spring, which is characterized by the spring constant, or in other words a force. Wouldn't it also be logical to use the force, i.e. the shear stress, for the correlation?

8.1.6.2 Calculation of the Shear Stress

The driving force (F) in the pump is the integrated spring [68]. Therefore it is important to consider the shear stresses in the annular gap, which are calculated as follows:

$$\tau = \frac{2 \cdot F \cdot h \cdot 10^6}{L \cdot \pi \cdot \left(d_A^2 - d_I^2\right)} \qquad \begin{aligned} F &= \text{Spring force} \\ h &= \text{Ball stroke} \\ L &= \text{Gap length} \\ D &= \text{Diameter (external/internal)} \end{aligned} \tag{8.4}$$

The formula can be used to derive relatively quickly the magnitude of the force of the steel spring used and from this the resulting shear stress. Shear stresses larger than 100 Pa can already be attained with a spring force of $F = 1$ N.

In Fig. 8.13 the viscosity is plotted vs. the shear stress for five products in the relevant range using data from shear stress ramp measurements.

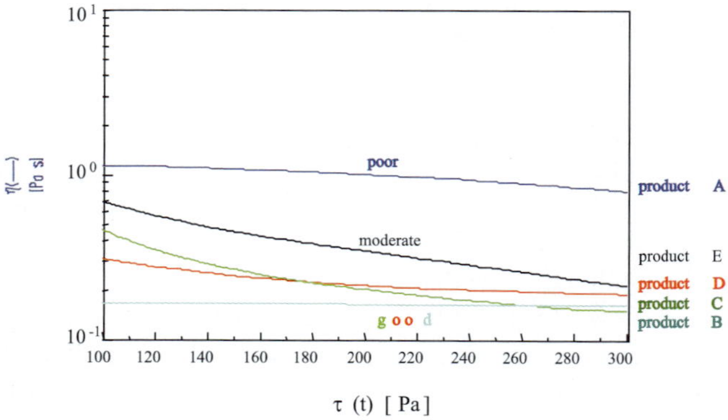

Fig. 8.13. Correlation of the viscosity data at $\tau = 200$ Pa with the pumping behavior

A clear result is obtained. The higher the viscosity in the shear stress range ($150 < \tau > 250$ Pa), the worse the pumping behavior. This can be explained by the film width that can build up on the container walls or remain adhered until the critical shear stress at the yield point is reached. The film can be thicker the higher the yield point or shear stress at the yield point. If the critical shear stress is exceeded by the mass of the emulsion film, the emulsion will start to slip down the container walls.

Since Newtonian products have no yield point they cannot build up a significant film on the container wall and hence exhibit a good residue emptying behavior (Product B). Unlike Product B, all other emulsions show a pseudoplastic flow behavior, resulting in a decrease in the viscosity at high shear rates or shear stresses. This can result in a viscosity that is lower than that of Newtonian products. It can be seen in Fig. 8.13 that the viscosity of Product A in the high shear stress range is clearly higher than that of the other products. This sample was assessed with poor pumpability in the pumping tests.

8.1.7 Stability Studies Using Yield Stress Measurements

The question of the stability of emulsions is always a concern in development and production. That was the starting point for studies on cosmetic emulsions using yield stress measurements. The aim was to determine [78] whether it is possible to detect stable and unstable emulsions based on test series. For these experiments stable and unstable formulations consisting of different emulsion systems were prepared. The samples were stored at room temperature (RT) and the measurements performed at $T = 25\,°C$ on unstressed samples. The emulsions were tested on the day of preparation (Day 0) and on Days 1, 2, 7, 21, 35 and 49 and the results compared with those of conventional standard tests.

When these measurements are repeated at discrete time intervals, information can be obtained on the stability of the sample at constant temperature based on

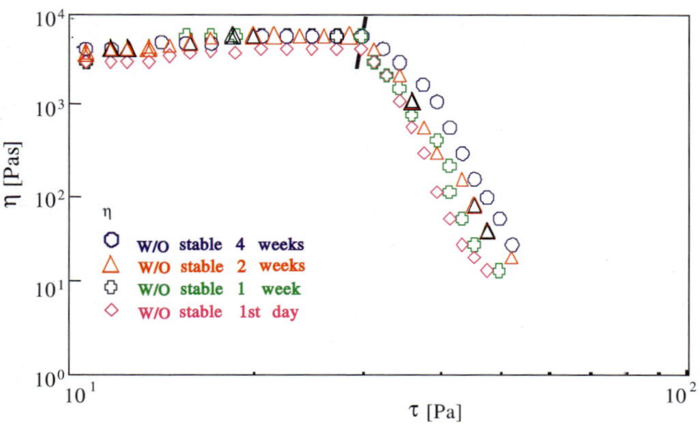

Fig. 8.14. Critical shear stress after storage

the change in the critical shear stress. The critical shear stress measurement for stable samples is time independent (Fig. 8.14); for unstable samples the apparent yield stress is shifted to higher values (Fig. 8.15).

However, certain factors need to be considered. Samples stored at RT and measured at $T = 25\,°C$ may appear to be stable in this test because they are only unstable at a higher or lower temperature than that used in this test.

Therefore only the emulsions that tend to be unstable at $T = 25\,°C$ will be identified with this test. Interestingly, it could be shown in this test that emulsions need a finite time after preparation before they attain a kind of resting state. This time is generally referred to as the maturation time. With the aid of the yield stress test it was possible to demonstrate that this maturation time is product dependent.

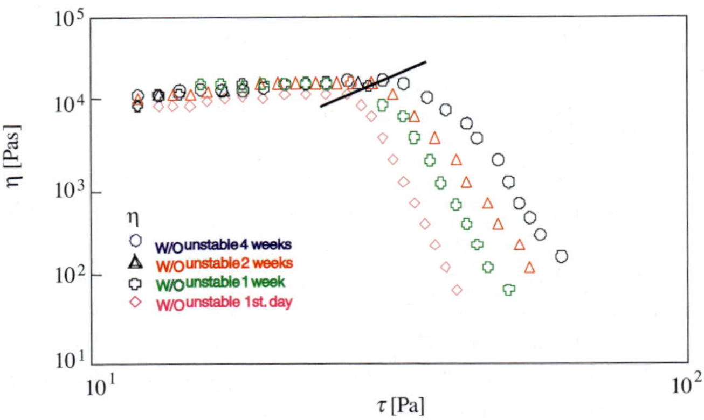

Fig. 8.15. Change in the critical shear stress after storage

8.1.8 Results Obtained

1. The yield stress can be used to detect the maturation process of emulsions. The exact duration could be determined by daily measurements with an automated measuring instrument. The time frame for this should be not more than 7 days for lotions and not more than 21 days for creams.
2. The yield stress determination at 25 °C cannot be used to obtain information on the instability of o/w emulsions. Yield stress measurements would need to be performed at temperatures at which instabilities arise.
3. Comparison of yield stress curves from yield stress measurements of w/o emulsions provides information on instabilities of creams.
4. No information on w/o lotions can be obtained because instabilities arise only with storage at 40 °C.

8.2 Steady Flow

Now that we have looked in detail at the stress ramp test and possible correlations of the yield points obtained from it we will consider the steady measurement. Unlike the shear stress ramp in which an integral viscosity is determined, under steady, meaning equilibrium conditions, viscosity data are measured for exactly one shear rate or shear stress. The measurement itself is not influenced by any other external factors. These measurements are therefore performed at constant temperature and shear rate and plotted against time. You are probably now wondering why we need steady measurements. This will be answered in the following chapter.

8.2.1 Determination of the Measuring Time

In a steady measurement the stress step described in Chap. 4 is often programmed. The temperature is kept constant and usually the test starts at a low shear rate. This shear rate is recorded for a certain time before the jump is made to the next shear rate level.

If the viscosity curves for all shear rates are plotted in one diagram vs. the time Fig. 8.16a is obtained. It is striking that the shear stress τ always shows a maximum [79] at the start of a new shear rate level and then gradually decreases to a constant level.

This overshoot depends on both the instrument and the sample. A measuring instrument needs a finite time to reach the desired pre-programmed value. As a result, the preset nominal value may be exceeded for a short while. However, the sample itself also needs time to reach constant conditions. The following observation is extremely important. The lower the shear rate is, the longer it will take for steady flow to be attained.

This has other consequences for our sample. As we already learned in Sect. 7.7, the samples to be measured differ the most distinctly at low shear rates and in the

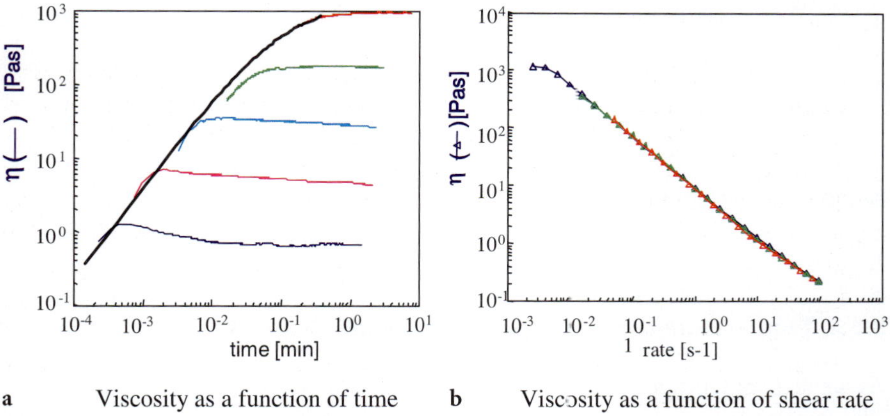

Fig. 8.16. **a** Viscosity as a function of time. **b** Viscosity as a function of shear rate

long measuring times associated with them. In some cases this can mean that it will be necessary to wait 5 min or more before taking a measurement. Since we want to record a flow curve, we will need several measuring points, and measuring times of 1 h can quickly accumulate for a complete flow curve. If we place an envelope curve over all measuring curves in Fig. 8.16 a and project it on the y-axis, we obtain the viscosity curve vs. the shear rate, of course only if the measuring times are long enough.

Another complication is the known fact that cosmetic emulsions contain water, emulsifiers, oils and even small amounts of fragrance. Fragrances especially are known to be highly volatile, but also oil and water can evaporate particularly at higher temperatures.

The onset of evaporation is shown by a more or less sudden viscosity increase that would not have been expected. This time-dependent measurement allows the rheologist to precisely define the boundary conditions for a measurement. In the example in Fig. 8.17 evaporation begins after 180 s. This means that the measuring points need to be recorded in 3 min. Afterwards the measuring instrument must be refilled. In the worst case this can mean especially for a low shear rate that just a single measuring point can be recorded within this time. This clearly shows that a viscosity curve cannot always and will usually not be obtained with a single filling. This in turn requires a large staff capacity of appropriately trained employees. Use of a suitably designed dome can considerably prolong the measuring times before a viscosity increase is observed.

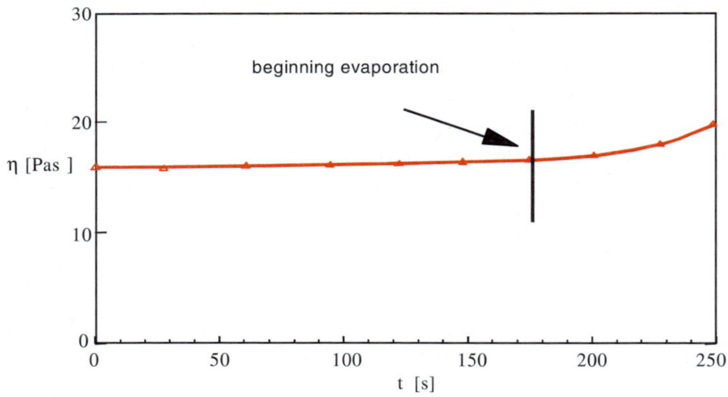

Fig. 8.17. Onset of evaporation of an emulsion at $T = 30\,°C$

8.2.2 Temperature Dependence of the Dynamic Viscosity

As we just said, evaporation must be expected especially at higher temperatures. The temperature dependence itself is, however, product specific. In Fig. 8.18 the results of measurements of an emulsion are plotted for three different temperatures.

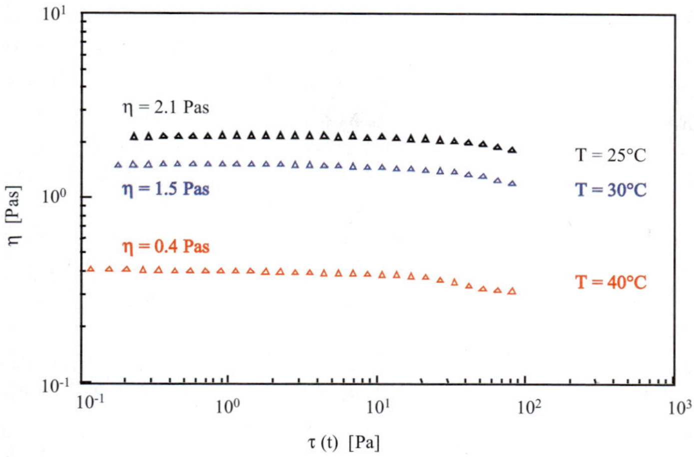

Fig. 8.18. Temperature dependence of a cosmetic emulsion

Apart from boundary conditions such as the maximum measuring time, the temperature is the most important parameter with the greatest influence on the measuring result. Some products exhibit extreme temperature dependence. Then it is essential to pay attention to the absolute measuring temperature.

8.2.3 Secondary Skin Feel

Steady measurements can also be used to determine the secondary skin feel [67]. As already mentioned in Sect. 8.1.1, the primary skin feel is the sensation occurring when an emulsion is initially applied to the skin. This is associated with small forces needed to make the emulsion flow. In the further course of application the product is subjected to high shear gradients because the film thickness of the cream on the skin decreases with increasing application time. The film thickness of the cream on the skin ultimately approaches the range of the diameter of the emulsified droplets. As a result oil is also rubbed directly on the skin. The sensory perceptions in this flow range where the product is almost completely spread on the skin comprise the term secondary skin feel.

Cosmetic emulsions of the w/o type can be stabilized by the addition of higher-melting waxes. Unlike fats and oils, waxes are not triglycerides but esters of higher primary monovalent alcohols with fatty acids. The waxes added to the products in the production process liquefy at temperatures of 70 to 80 °C, mix with the continuous oil phase and crystallize on cooling. In this way a solid structure arises similar to that of the polymer network of an o/w emulsion. The disperse water phase of the w/o emulsion is also mechanically demobilized. Therefore the large viscosity changes occurring with o/w emulsions during application to the skin do not take place because the stabilized wax structure is destroyed more slowly. The difference between the primary and secondary skin feel is therefore smaller.

8.2.3.1 Investigation of the Secondary Skin Feel

For correlation of the secondary skin feel with the rheological variables, the sensory assessment of the products is compared with their steady viscosity curves for shear rates up to $\dot{\gamma} = 10^5\,\text{s}^{-1}$. This maximum rate of shear is estimated assuming that the spreading velocity is $v = 1\,\text{m/s}$ and the film thickness of the cream $x = 0.01\,\text{mm}$:

$$\dot{\gamma} \equiv -\frac{dv}{dx} \approx \frac{\Delta v}{\Delta x} = \frac{1\,\text{m/s}}{10^{-5}\,\text{m}} = 10^5\,\text{s}^{-1} \tag{8.5}$$

Stern [62] cites shear rate estimates by different authors in the range $10^4\,\text{s}^{-1} \leq \dot{\gamma} \leq 10^5\,\text{s}^{-1}$. Shear rates of up to $2500\,\text{s}^{-1}$ can be achieved with the DSR and RDA rotary rheometers from Rheometric Scientific. To obtain higher shear rates of $10^4\,\text{s}^{-1} \leq \dot{\gamma} \leq 10^5\,\text{s}^{-1}$, the Rheomat 2000 high-pressure capillary viscometer (HSCV) from Göttfert must be used.

The study on the correlation of the secondary skin feel and the viscosity curve was performed with the same products as those used for the study on the primary skin feel.

To determine experimental values for the shear rate occurring on application of a cosmetic emulsion to the skin, the viscosity of a series of Newtonian oils was measured and the corresponding skin feel determined by a test panel. The aim was to determine the viscosity of the oil considered to have the optimal skin feel.

Since the optimal viscosity was determined using Newtonian oils, the absolute value did not change over the whole range of shear rates studied. After the viscosity curves of the products were measured, the rate of shear was determined at the point where the viscosity reached the previously determined optimal value. The shear rate determined by this method was then correlated with the results of the sensory panel test.

In Fig. 8.19, the results measured with the different instruments are shown with the confidence interval (95%). Oils C to G could not be measured with a high performance capillary viscometer (HCV) because of the low viscosity. On the one hand, the accuracy of the available pressure transducer was insufficient for the pressure differences to be measured ($\Delta p < 0.5$) and on the other hand, the oils flowed through the capillary under the force of their own weight alone. Due to the better reproducibility only the values measured with the DSR are presented in Table 8.5.

Oil D, which had a viscosity of $\eta = 0.024\,\text{Pas}$, received the best assessment and oils C ($\eta = 0.036\,\text{Pas}$) and F ($\eta = 0.0064\,\text{Pas}$) a somewhat poorer assessment. It is evident that the absorption capacity perceptible on the skin increases with decreasing viscosity. The oils can be clearly distinguished in the sensory test. Since most cosmetic emulsions show non-Newtonian flow behavior, it is possible to find a rate of shear at which the viscosity is $\eta = 0.028 \pm 0.005\,\text{Pas}$. This shear rate is approximately $\dot{\gamma} = 5000\,\text{s}^{-1}$ for the o/w lotions and approximately $\dot{\gamma} = 500\,\text{s}^{-1}$ for w/o/w creams. The shear rates measured by this method are clearly lower than the estimated value of $\dot{\gamma} = 10^5\,\text{s}^{-1}$. This is due to the dependence of the shear rate of product application on the type of product as well as to the fact that the sensory

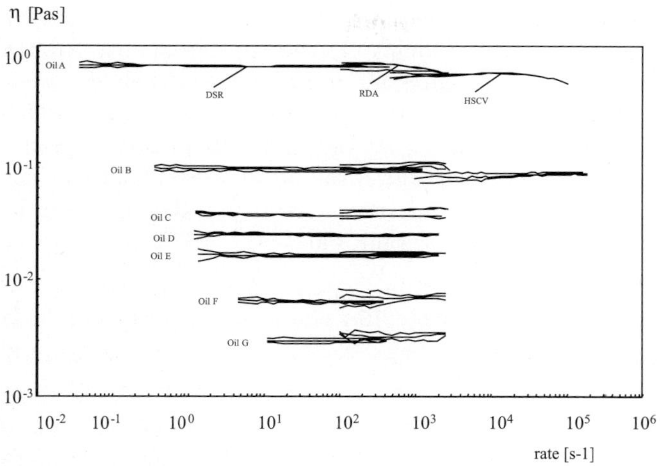

Fig. 8.19. Reproducibility of viscosity measurements of Oils A–G

Table 8.5. Different oils and oil mixtures

	Components and composition in parts by volume	Viscosity η [Pas]	Verbal sensory assessments of skin feel
Oil A	Oil1 = 100%	0.674	No assessment
Oil B	Oil1/Oil2 = 50%/50%	0.0878	"Oily, greasy, neutral skin feel"
Oil C	Oil1/Oil3 = 50%/50%	0.0357	"Spreads well, somewhat oily and dull"
Oil D	Oil1/Oil2/Oil3 = 33%/33%/33%	0.0242	"Somewhat oily, disappears slowly, very good skin feel"
Oil E	Oil2 = 100%	0.0161	No assessment
Oil F	Oil2/Oil3 = 50%/50%	0.00643	"Watery, dry skin feel"
Oil G	Oil3 = 100%	0.00298	No assessment

skin feel is product-specific. This is understandable if one considers how each type of product is used. A lotion is applied to large areas of the skin like the arms, legs, and trunk. A cream is usually applied to a smaller area such as the face and rubbed in with a lower shear rate than a lotion.

8.3 Oscillatory Measurements

Mechanical oscillatory measurements allow determination of rheological material functions (Sect. 5.10). The sample to be tested is subjected to a sinusoidal deformation and the complex shear modulus G* can be calculated from the response to the oscillatory load.

Mathematical separation into a real and imaginary part provides a measure of the stored elastic energy (storage modulus G') and the energy lost through viscous flow (loss modulus G''). If substances are studied in the viscoelastic range, oscillatory measurements offer a nearly non-destructive measuring method which allows structural studies on complex systems. Studies on the viscoelastic properties of polymer fluids have long been known [80]. It can be seen from the frequency dependence of the storage modulus G' for such systems that the curve can be divided into four zones in the double logarithmic plot.

IV Glassy region Starting at the far right in Fig. 8.20. At high frequencies (or correspondingly at low temperatures) the mobility of the chain segments is highly restricted. The vibrational energy is stored by deformation of the bond angle and distances. A plateau is reached, the glassy region.

III Transition region With decreasing frequency (increasing temperature) the mobility of the chain segments slowly increases. Consequently, the storage modulus G' decreases. There is a transition here from the glass region to the rubbery-elastic region.

II Plateau region In this region the mobility of the polymer chains continues to increase with decreasing frequency, but the polymer chains still do not slip past each other. A network forms with semi-solid entanglement points because the shear rate is greater than the relaxation time τ_n. This region can be explained by the theory of rubber elasticity and offers the possibility of calculating the entanglement molecular weight Me, the network density ν_e and the crosslink interval ξ using the storage modulus in the plateau region G'_p.

I Flow region With an even lower frequency the deformation energy can no longer be stored and the storage modulus G' decreases further. In this region the system tends to dissipate more of the input energy through viscous flow with decreasing frequency. This region is characterized by the relaxation time τ_n from the reptation concept. At these frequencies the polymer chains can slip past each other through reptation motions. Figure 8.20 shows the storage modulus G'

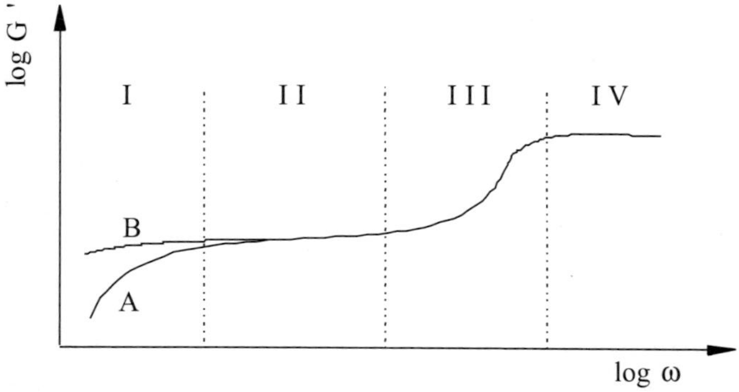

Fig. 8.20. Graph of the frequency dependence of the shear modulus G' according to Kulicke

as a function of the angular frequency ω. The regions described vary strongly depending on the polymer fluid. Linear and branched polymers follow *Curve A*, whereas mainly crosslinked polymers (gels) follow *Curve B*. There is no flow region for these systems. Short-chain, linear polymers that cannot form a network exhibit a continuous transition from flow region I to glass region IV with no distinct plateau region II. The width of the plateau region for polymer fluids depends on the molecular weight as well as the molecular weight distribution.

A similar picture is obtained when cosmetic emulsions are observed in a correspondingly large frequency window, assuming a linear viscoelastic range.

For the emulsions studied *regions I* to *III* can be detected (Fig. 8.21), although the systems contain no polymers. The emulsifiers used have molecular weights of 780 and 3100 g/mol respectively. The results of polymer analysis can be transferred to emulsions if we assume the dispersed water droplets slip past each other like the reptation motion of polymer chains.

Most emulsions have an internal network structure resulting from intermolecular interactive forces. When a stable sample is stressed in the linear viscoelastic range the storage modulus G' predominates and is usually larger than the loss modulus G'' by a factor of 10 to 1000 (Fig. 8.22). For stable samples the curves for both moduli are nearly parallel over the entire frequency range measured, with a slight increase in the slope at high frequencies.

No conclusions about this internal network can be derived from the complex dynamic viscosity η^* (see Figs. 8.22 and 8.23) calculated for each sample, as it decreases, as would be expected, with increasing frequency in both cases.

In Fig. 8.24 the reproducibility of the frequency measuring method is illustrated using the example of an o/w lotion that is $< 5\%$ over the entire frequency range. Interestingly, G' and G'' seem to intersect at very low frequencies for this sample. If the absolute values of the moduli G' and G'' in the preceding figures are now examined, a correlation with the internal network becomes clearly apparent and it can be determined that specific formulations also show characteristic curves.

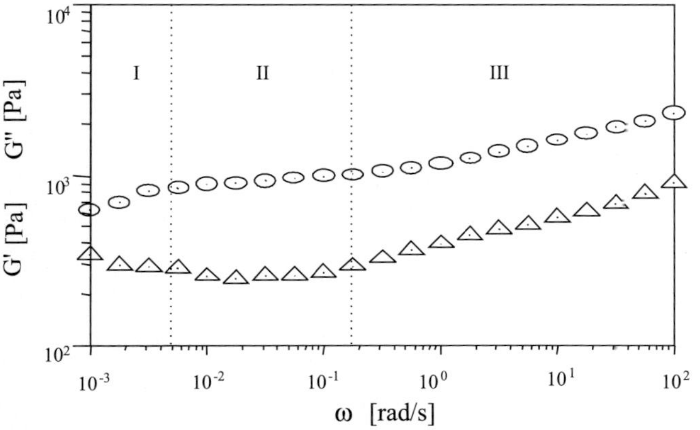

Fig. 8.21. Frequency dependence of the storage and loss modulus of a w/o emulsion

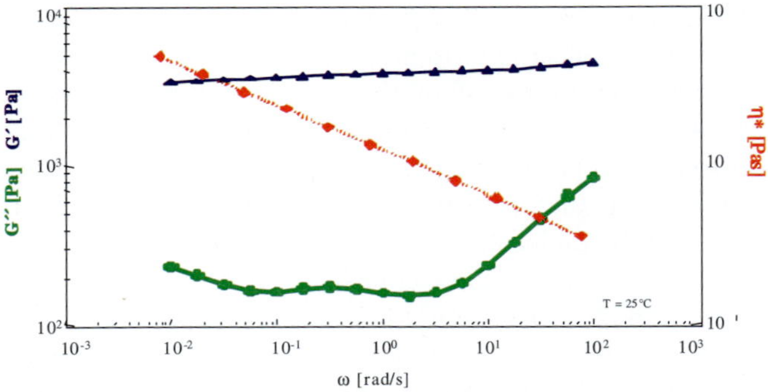

Fig. 8.22. Frequency dependence of a w/o cream with greater internal strength

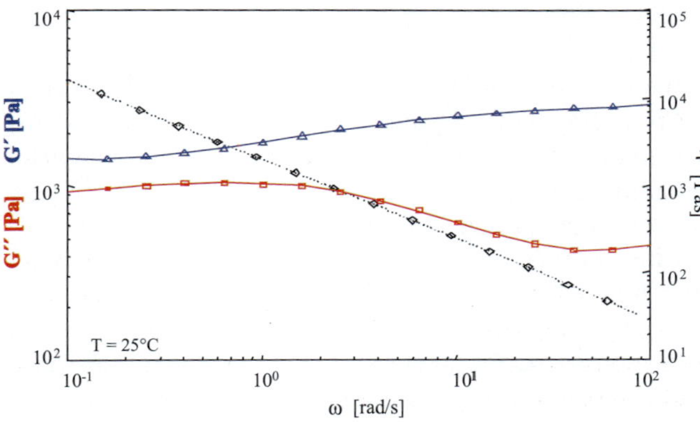

Fig. 8.23. Frequency dependence of an o/w cream with moderate structural strength

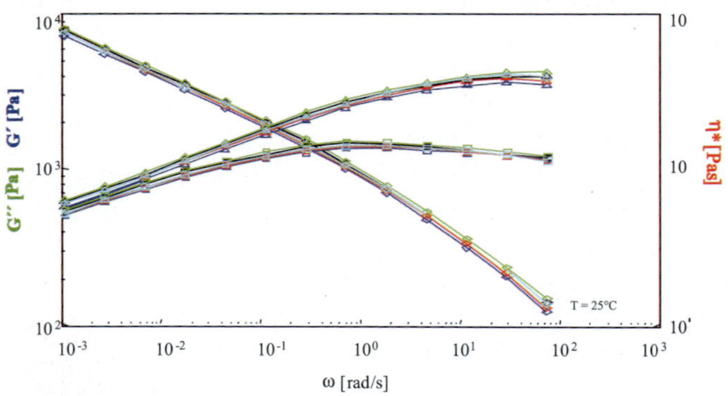

Fig. 8.24. Frequency dependence of an o/w formulation with little network structure

As already indicated in Fig. 8.24, there are formulations for which the storage modulus G' and the loss modulus G'' intersect; this is known as crossover (Fig. 8.25). At low frequencies the viscous part is greater than the elastic part; the sample

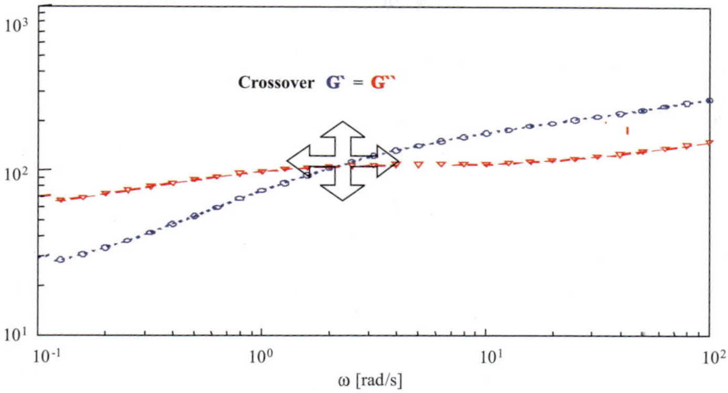

Fig. 8.25. Frequency dependence of a hydro dispersion gel with crossover

behaves like a liquid. With increasing frequency both the loss modulus G'' and the storage modulus G' increase, with the storage modulus increasing clearly faster. Above a certain frequency characteristic for this product the storage modulus G' is larger than the loss modulus G''.

This means that the internal structures can no longer withstand the marked deformation occurring at this and higher frequencies. This characteristic property can be used to regulate the production process for this frequency-dependent measurement is very quick (< 15 min) and deviations of the frequency or the crossover modulus G_{cross} allow conclusions about the process to be drawn.

The clearly greater informative value of the moduli, unlike that of the yield point results or the viscosity curve is of crucial importance especially in the area of research and development. For instance, information on the effect of different raw materials on the base formula can be obtained in a very short time from the moduli determination.

In the next example we will examine the effect of homogenization during the production of cosmetic emulsions and its consequences for the rheological properties. As we already learned in Sect. 4.6, smaller droplets are formed during homogenization. To demonstrate the influence on the emulsion of the energy input associated with this we took samples at defined intervals during homogenization and subjected them to a frequency test.

In the plot of tan δ vs. the angular frequency in Fig. 8.26, it can easily be recognized that at low frequencies only small differences can be detected between the individual samples. Clear differences become apparent only at high frequencies. What is the explanation? At the start of the homogenization process our emulsion has a semi-solid consistency. With increasing energy input smaller droplets are generated and an internal network begins to form. Consequently, the loss factor tan δ decreases with a higher degree of homogenization. The mobility of individual droplets is described best at high frequencies because a few small droplets influence each other less than many small ones. The intercept with $\tan\delta = 1$ is a possible parameter for discrimination of the individual homogenization steps because it is

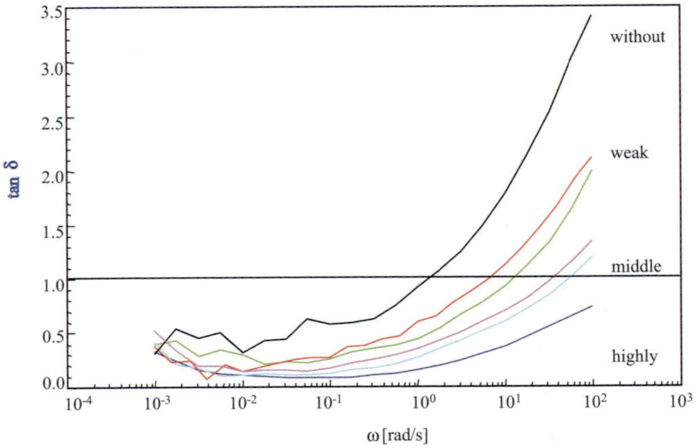

Fig. 8.26. Emulsions of varying degrees of homogenization

known that this is the point at which $G' = G''$. In addition, this point can be easily determined.

8.3.1 Temperature Dependence of the Moduli

To determine the temperature dependence of the moduli the frequency, amplitude and temperature-time profile (heating or cooling rate) are kept constant [65]. These experiments are used to study the freezing, softening and melting behavior. It is important to determine the correct frequency and linear viscoelastic range (LVR) for the temperature range to be measured. From frequency measurements we learned that the moduli run nearly parallel with a slight increase towards higher frequencies. This fact as well as the wish to characterize samples in the resting state necessitate use of low frequencies. This in turn requires long measuring times. As a compromise, the angular frequency $\omega = 1$ rad/s is often used. This also has the advantage that the complex shear modulus G^* determined is also the complex dynamic viscosity η^* (see Sect. 5.11.5). To obtain the correct LVR the amplitude must be determined in amplitude tests conducted at least for the starting and final temperature.

Now we will measure an o/w emulsion, starting with a temperature measurement at 25 °C and cooling 1 K/min to −50 °C (LVB and $\omega = 1$ rad/s).

From 25 °C the storage modulus G' as well as the loss modulus G'' increase slowly in parallel. At $T = -8$ °C both moduli suddenly shoot upwards by several powers of ten (increase). What's happening to the sample? Remember that the external disperse phase in an o/w emulsion is water. What happens to water when it cools? At some point it freezes. This test allows the determination of the freezing point of an o/w emulsion (Fig. 8.27). Since this is a phase transition from a solid to liquid state, we speak of a first order phase transition.

At this point we need to ask whether the value measured for the freezing point depends on the cooling rate. To answer this question the cooling rate was

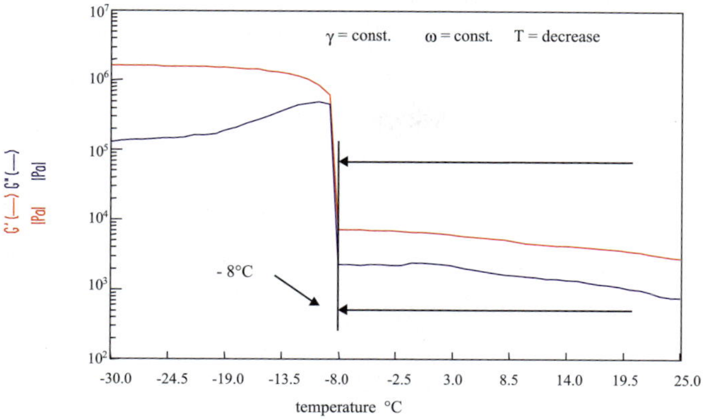

Fig. 8.27. Cooling curve for an o/w emulsion

varied from 1 K/min to 10 K/min and the freezing point determined for another o/w emulsion. The results are presented in Fig. 8.28.

It can be seen from the graph that for cooling rates between 1 and 5 K/min the measured freezing point remains the same at $T_f = -12\,°C$. A lower freezing point is measured only with even faster cooling rates. This can be explained by the fact that the sample continues to cool during the measurement and does this so quickly at a cooling rate of 10 K/min that it still has not frozen when the measurement at $T = -12\,°C$ begins. The whole phenomenon is a time effect. What really matters for determination of the freezing point is that cooling rates < 5 K/min have no influence on the measured result.

Now the question arises: what is the freezing behavior of a w/o emulsion? To answer this question we chose the same test conditions and began cooling a w/o cream from 25 °C at a rate of 1 K/min. The results are presented in Fig. 8.29.

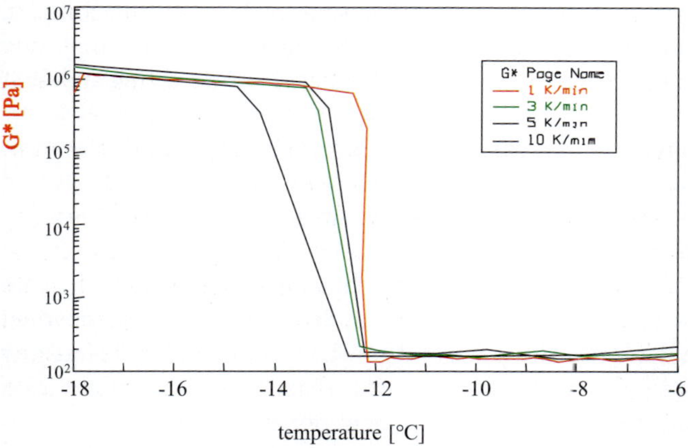

Fig. 8.28. Variation of the cooling rate for determination of the freezing point

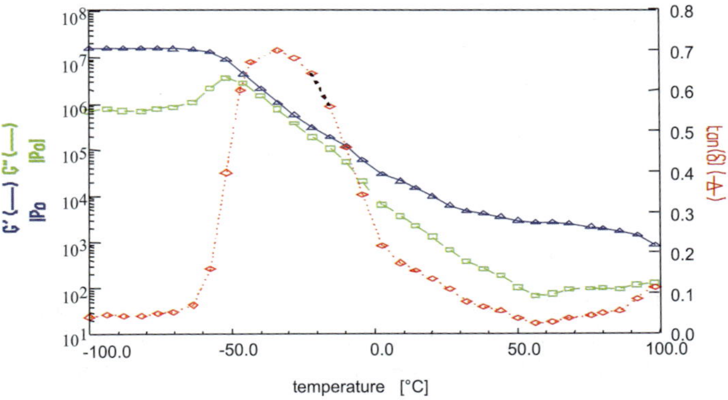

Fig. 8.29. Complex dynamic glass transition of a w/o emulsion

The cooling curve of a w/o emulsion (Fig. 8.29) differs significantly from that of an o/w emulsion (Fig. 8.28). The moduli do not increase spontaneously by several powers of ten, but they are distinctly higher in the cold with absolute values that indicate solid properties but a transition that is not as rapid. We see behavior similar to that known for polymers when they are cooled. One speaks of a complex dynamic glass transition region Tg dyn. Either the maximum of the loss modulus G'' (the absolute value of G'' is closer to the results of a DSC study) or the maximum of the loss factor tan δ is evaluated for the familiar reason of improved reproducibility because the influence of geometry is eliminated. The reason a w/o emulsion reaches this maximum is once again the external phase, which in this case is the oil. Several oils or oil mixtures are often used to produce emulsions. This is what is responsible for their polymer character. The result of the freezing point measurement of an emulsion can be used to determine whether the formulation is a w/o or o/w emulsion. This is a costly analysis considering that we can determine this quicker and cheaper simply by applying an emulsion to our hands and washing it off with water. An o/w emulsion can be easily washed off, whereas a w/o emulsion leaves behind an oily, fatty residue.

The freezing point determined for o/w emulsions is useful in two ways. It provides information on the cold storage stability of the emulsion because the lower the freezing point is the less sensitive the sample is to the cold. However, freezing of the external phase can also affect the emulsion properties measured subsequently at room temperature. A notable example is the higher viscosity measured after cold storage.

To study this phenomenon, we divided each of four o/w emulsions in half. One half was stored for one week at $T = -10\,°\mathrm{C}$. In the same period we recorded the viscosity function curve of the other half at $T = 25\,°\mathrm{C}$ and subsequently determined the freezing point. After cold storage the samples were stored for 4 h at room temperature before their viscosity curves were recorded at $T = 25\,°\mathrm{C}$.

Compiled in the table are the difference in viscosities measured at $\gamma = 10\,\mathrm{s}^{-1}$ before and after cold storage and the rheologically measured freezing point. It can

Table 8.6. Viscosity increase after cold storage

	Emulsion A	Emulsion C	Emulsion D	Emulsion B
Viscosity difference $\Delta\eta$	No difference	2400 mPas	1900 mPas	1500 mPas
Freezing point T_{frozen}	−12 °C	−4 °C	−6 °C	−7 °C

be seen that the viscosity of Emulsion A, which had a freezing point of $T_{\text{cool.}} = -12\,°\text{C}$ and therefore was much lower than the storage temperature, did not change after cold storage. In contrast, an interesting relationship is apparent between the freezing point and the change in viscosity for the other emulsions: the higher the freezing point the greater the viscosity difference (Ranking Emulsion B, D, C). SEM micrographs of Sample B taken before and after cold storage provide an explanation for this phenomenon.

In the picture of the emulsion before cold storage on the left (Fig. 8.30), a distinct oil phase (A) can be clearly recognized embedded in the water phase (B). In the picture on the right the changes occurring after one week of cold storage at $T = -10\,°\text{C}$ can be seen. The oil phase A is dispersed much more finely in the water phase and small ice crystals C can even be recognized. These originate from the freezing process at $T = -10\,°\text{C}$. Water strives to form pure ice crystals. As a result, soluble and insoluble ingredients accumulate at the interface and are compressed, forming a new three-dimensional structure. These structural changes (ice crystals and interfacial structures) influence the physical properties of samples.

This type of test method will, however, only be useful if such changes can be measured quickly and made immediately available to the developer as a tool for product optimization. To illustrate the usefulness of this method we studied the effects on the freezing point of adding small amounts of an alcohol to this type of emulsion.

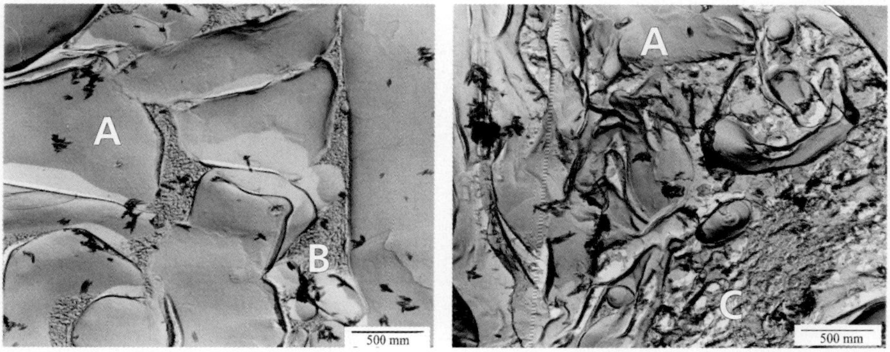

Fig. 8.30. TEM Micrographs of Emulsion B before (*left*) and after (*right*) cold storage

Table 8.7. Freezing point depression with alcohol and its effect on viscosity

	Orig. emulsion	Emulsion +0.2% alcohol	Emulsion +0.4% alcohol
Viscosity difference	2.5	2.1	1.9
Freezing point Tg_{dyn}	−7 °C	−9 °C	−11 °C

As would be expected, addition of small amounts of alcohol lowered the freezing point. At the same time the viscosity increase after cold storage at $T = -10\,°C$ was reduced from a factor of 2.5 to 1.9. This is just one small example of how rheology measurements can be a useful tool in development.

8.3.2 Temperature Stability

Now that we have subjected the sample to a cooling process and can describe the freezing behavior it would seen logical to investigate what happens to the sample when it is heated [65]. We will use the same temperature program for heating as for cooling, 1 K/min, starting once again at $T = 25\,°C$. It can be seen in Fig. 8.31 that the curves for the storage and loss moduli are nearly parallel until $T = 50\,°C$. Then there is a decrease in both moduli, with the storage modulus G' falling much more rapidly than the loss modulus G''. In other words the sample becomes increasingly less viscous.

If several emulsions are measured with this test, a specific softening point is obtained for each product, as can be seen in Fig. 8.31. The reproducibility of the measured result depends on the temperature steps alone and is $T_{rep} = \pm 1\,°C$ for a heating rate of $\Delta T = 1\,K/min$

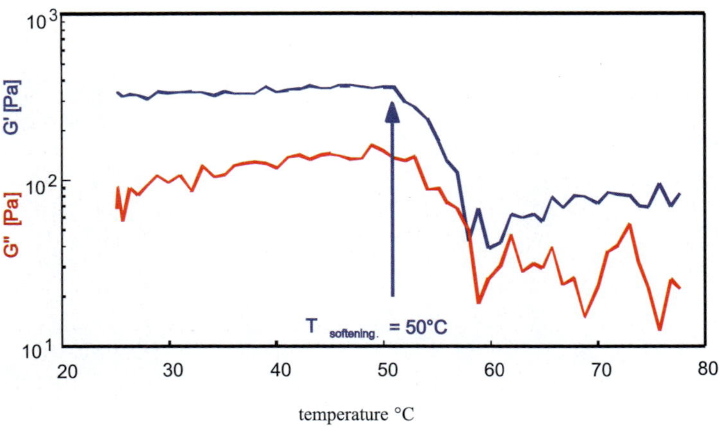

Fig. 8.31. Determination of the temperature-dependent onset of flow of an emulsion

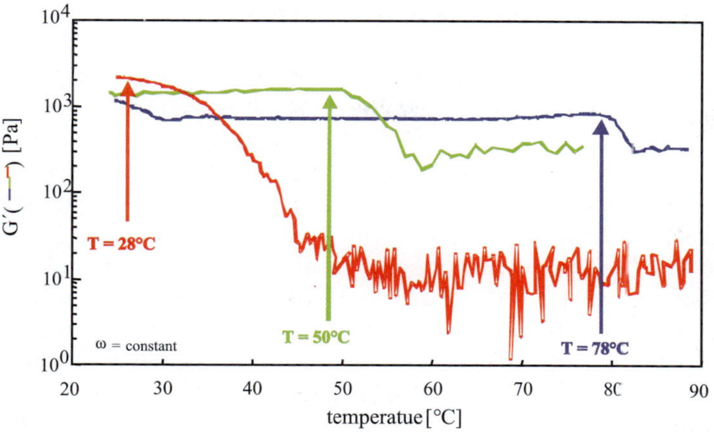

Fig. 8.32. Determination of the softening point of different emulsions

The differences can be fairly large. For instance, one product can already have a softening point at $T = 28\,°\mathrm{C}$ and another not until $T = 78\,°\mathrm{C}$. When combined with the cooling test, information on the temperature behavior can be obtained for any sample. Using these two measurements we can determine the temperature limits for every sample between which the moduli G' und G'' remain nearly unchanged and run almost parallel. Consequently, the temperature range is determined in which the sample is independent of the temperature and can therefore be considered stable (based on the temperature limits determined).

Instabilities are not only temperature dependent (see yield stress measurements for stability), but also time dependent. And the two dependencies are interlinked. Using the dynamic mechanical swing test it is possible to scan a sample several times with a defined temperature profile (Fig. 8.33) (cooling with a defined slow rate and then heating) with virtually no destruction occurring.

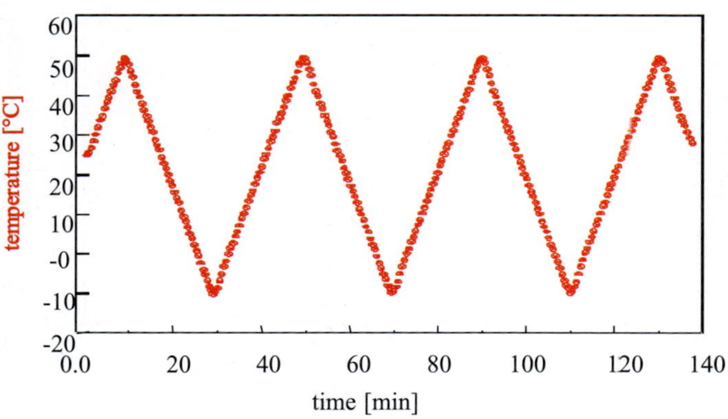

Fig. 8.33. Dynamic temperature swing test

8.3.3 Rheological Swing Test for Temperature Stability

Many tests can be created with results that need to be put together like pieces of a puzzle in order to be able to evaluate the stability. A temperature swing test [66] is an interesting experiment in terms of physical stress because by changing the temperature in a defined manner conditions much like those experienced in use by the consumer can be simulated. In practice, it has been shown that the developer of a cosmetic emulsion examines a temperature window in the range $-10\,°C < T > 50\,°C$. Therefore a dynamic swing test was developed that starts at $T = 25\,°C$. After heating to $T = 50\,°C$ with a constant temperature ramp, the sample is cooled to $T = -10\,°C$ with the same but negative temperature ramp. A sweep of this temperature profile is performed three times. Based on experience obtained from determination of the maturation time (see yield point measurement), these measurements were performed no earlier than 48 h after production. The result one would expect is:

A sample is classified as stable if the moduli G' and G'' are constant

- in the temperature range measured,
- over time and
- with changing stress (distance, temperature, frequency or time).

As we can see in Fig. 8.34, the result is not what we expected. In the cold the moduli always increase slightly, which can be explained by solidification of the emulsion. These high demands can only be met by an ideal sample. In reality, certain restrictions must be placed on the external experimental conditions such

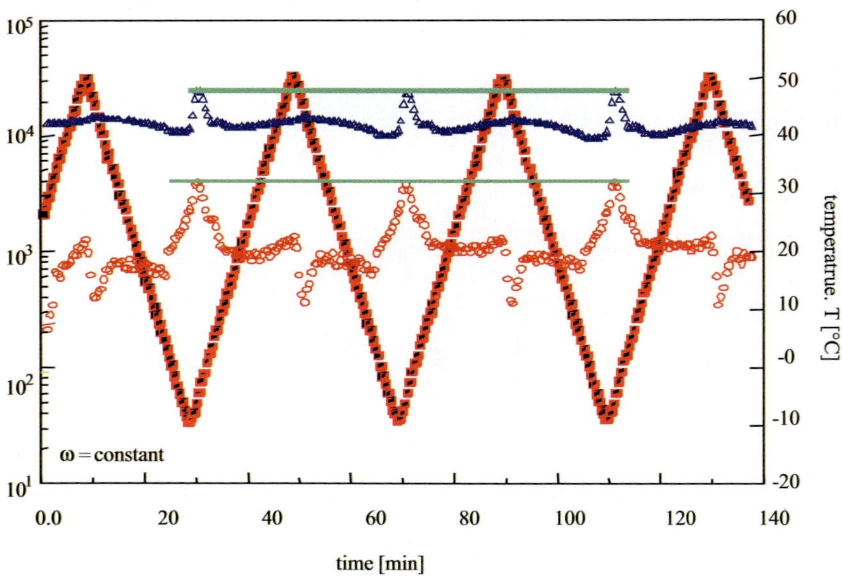

Fig. 8.34. Dynamic swing test for a stable w/o emulsion

as a temperature range of $-12\,°C$ to $+60\,°C$, or a frequency range of $\omega = 1\,rad/s$ to $\omega = 100\,rad/s$, or keeping the measuring time as short as possible $t < 12\,h$ etc. The experiments were performed using the Rheometrics Dynamic Analyser RDA and the Advanced Rheometric Expansion System ARES from Rheometric Scientific. However, this much can already be said: rheological measurements allow detection of temperature effects on the sample during the whole experiment and thus determination of critical temperatures that cause a change in the emulsion. Consequently, a new hypothesis must be formulated:

A sample is classified as stable if the moduli G' and G'' have identical values at recurring temperatures at constant frequency and amplitude (in the linear viscoelastic range) regardless of the number of temperature sweeps performed.

The temperature swing test for an unstable w/o emulsion is presented in Fig. 8.35. Changes in the moduli from the first cycle to subsequent cycles are clearly apparent. Let us now take a look at another plot of the same measurement, namely the plot of the loss factor $\tan\delta$ vs. the temperature.

In this plot (Fig. 8.36) the temperature behavior of the sample is now clearly recognizable. The loss factor is $\tan\delta < 0.6$ over the whole temperature range. Consequently this is an emulsion with a high internal strength or, formulating it more precisely, one that can be characterized by a pronounced network. The behavior in the cold is striking. In the range $-10\,°C < T < +10\,°C$ a kind of hysteresis curve can be seen that is shifted increasingly towards higher δ values from cycle to cycle. This change indicates that the sample studied has a slight stability problem in the cold. In the higher temperature range the curves are noisy but reproducible. This is due to the fact that in this temperature range the absolute values of G' and G'' decrease (i.e. the sample has begun to soften) but they decrease proportionately. This sample was stored at RT and re-measured

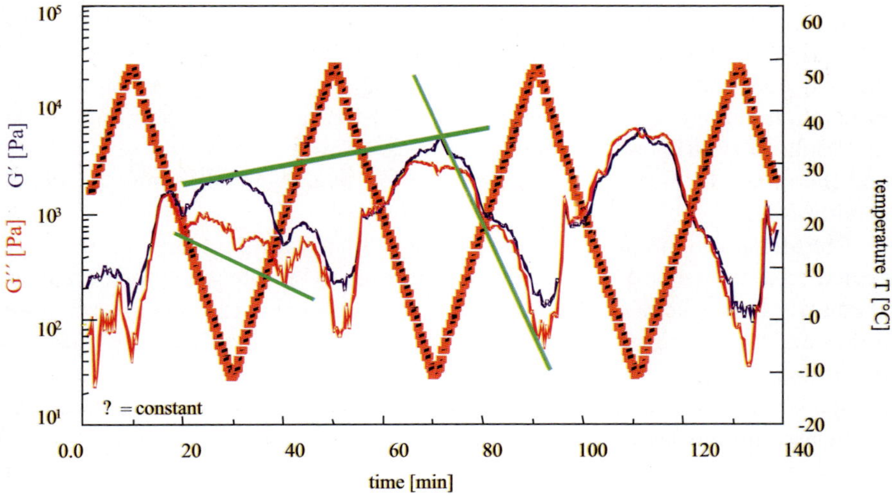

Fig. 8.35. Dynamic swing test of an unstable w/o emulsion

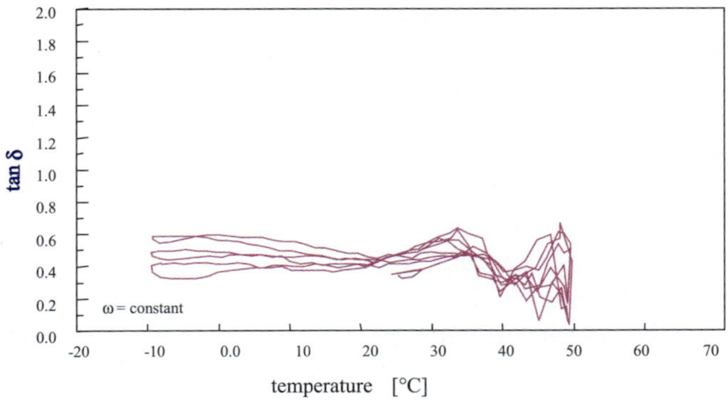

Fig. 8.36. Dynamic swing test of an unstable w/o emulsion

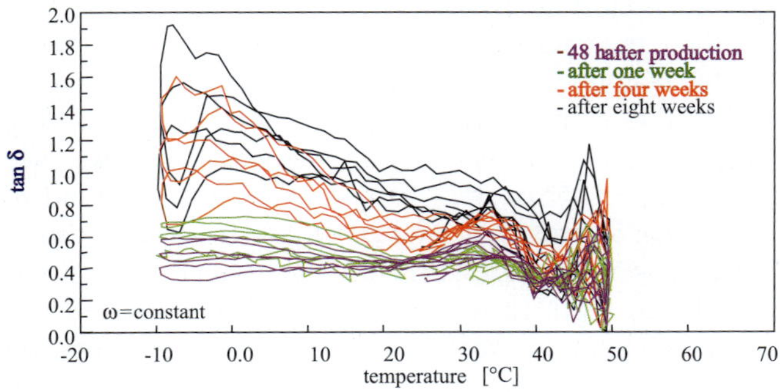

Fig. 8.37. Temperature swing test for an unstable w/o emulsion measured at weekly intervals

in the temperature swing test at weekly intervals. The result can be seen in Fig. 8.37.

Problems in the cold already discovered during the first measurement 48 h after production of the emulsion intensify in repeated measurements. Shown here are the results after 48 h, one, four and eight weeks. Now we need to do a crosscheck and measure and evaluate the stable emulsion from Fig. 8.38 under the same conditions.

The stable w/o emulsion does not change either with the temperature or repeated measurements. Therefore this test appears to be able to tell the developer at an early stage whether a new formulation has temperature problems and, more importantly, at which temperature problems will arise. This is helpful in that raw materials are characterized by corresponding measurements such as the melting point of a wax. This result of course allows no global statements on the correctness of the assumption because only two samples were measured and both were the same type of emulsion – w/o. Several w/o and o/w emulsions have since been characterized with this test and the results compared with results of standard tests.

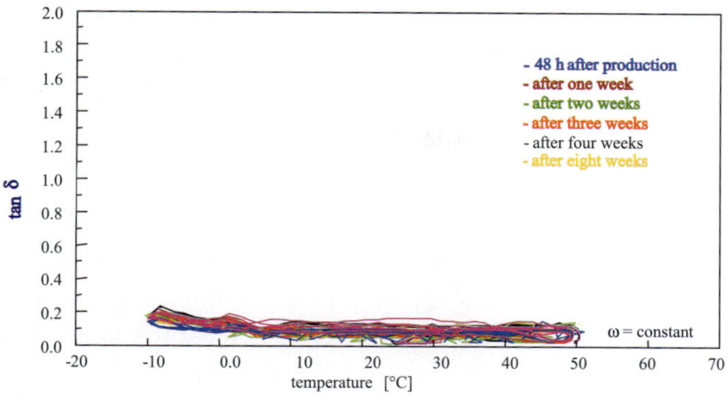

Fig. 8.38. Temperature swing test for a stable w/o emulsion measured at weekly intervals

Table 8.8. Comparison of cycle test results with those of standard storage tests

W/O emulsions n = 137						
Temperature swing test	Other visual tests					
48 h after production	< One week	< One month	< Half a year	< One year	< Two years	> Two years
37% Stable	Stable	Stable	Stable	Stable	Stable	Stable
28% Problem with minus temp.	Stable	14%	9%	5%	–	–
35% Problem in the heat	Stable	Stable	11%	14%	6%	4% Stable

W/O emulsions n = 178						
Temperature swing test	Other visual tests					
48 h after production	< One week	< One month	< Half a year	< One year	< Two years	> Two years
26% Stable	Stable	Stable	Stable	Stable	Stable	Stable
32% Problem with minus temp.	Stable	16%	9%	7%	–	–
42% Problem in the heat	Stable	Stable	12%	14%	8%	8% Stable

A total of 137 w/o emulsions and 178 o/w emulsions of a wide variety of formulas were freshly prepared and measured for the first time with the cycle test 48 h after preparation. At the same time standard tests such as storage at +40 °C, −12 °C etc. were performed and the results assessed visually. These parallel measurements were terminated after two years. A stable emulsion could already be identified reliably after 48 h regardless of whether it was the o/w or w/o type by plotting the cycle test results as moduli vs. time or $\tan d$ vs. temperature. All samples of unstable emulsions were also detected early as problem cases (hot or cold) by the cycle test. The same result was not obtained with the standard test sometimes until after weeks or years. Interestingly, 4% of the w/o emulsions were ranked as critical at high temperatures in the cycle test. For the o/w emulsions this was true for 8% of the samples. The cycle test therefore also reveals incipient heat problems that would not become evident during normal storage.

Dynamic mechanical thermal analysis (DMTA) allows virtually non-destructive measurement of the viscoelastic properties of cosmetic emulsions. The advantage of the method is that information can be obtained on the rheological properties under constant experimental conditions with relatively small sample amounts. The temperature swing test simulates application of a short temperature stress on emulsions. It resembles actual practice and allows continuous observation at any temperature. Significant, reproducible absolute values that correlate with common stability measurements are obtained as the result. Testing times for emulsions are considerably shortened (a few hours rather than weeks) and development times for new emulsions clearly reduced by control rheological measurements performed from the laboratory batch through pilot plant tests to production samples.

If it is already possible to measure an emulsion with the cycle test fairly soon after production, it must also be possible to use this method to optimize product formulas. Two samples were chosen to check this. One had a problem in the cold and the other at elevated temperatures.

The first measurement (*blue curve* at the top in Fig. 8.39) already revealed a problem in the cold. The formula was modified during development by adding a small amount of cetiol alcohol. After 48 h the modified sample was measured, giving the *green curves* in the middle. The formula was further modified by increasing the glycol content. This gave the optimized product variant on the second try.

In the second example (Fig. 8.40) a sample with very slight temperature problems above 30 °C could likewise be quickly optimized in collaboration with development by making a minor correction to the emulsifier composition.

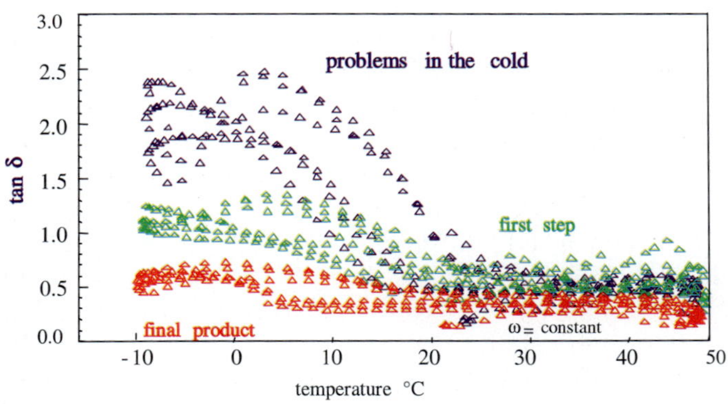

Fig. 8.39. Optimization of a sample with problems in the cold

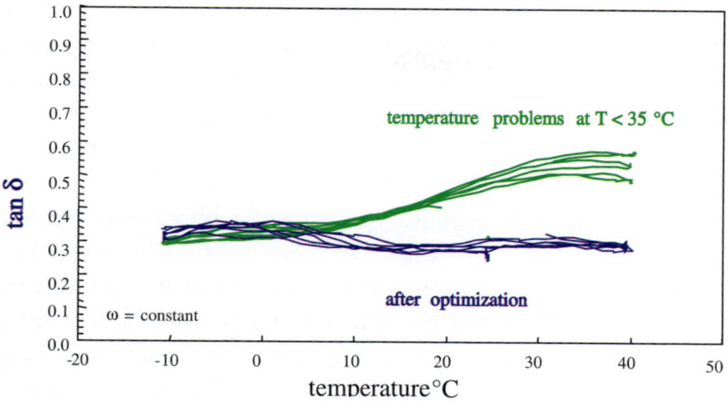

Fig. 8.40. Optimization of a sample with slight temperature problems above 30 °C

The cycle test is very well suited for optimization of formulas during development. The measurement can be performed after a relatively short time, namely 48 h after preparation. With many samples the developer must wait at least a week before a change can be seen. This test makes it possible to predict the temperature stability but not the storage stability although at first glance this might seem to be the case from the repeated measurements over weeks. The cycle test cannot replace long-term stability testing.

It is conceivable that it might be possible to make such statements on the basis of comparative measurements. A sample that was characterized in the cycle test as temperature resistant and also classified as stable long-term in conventional tests could be used as the reference. A new, unknown sample that shows poor behavior in the cycle test would accordingly be ranked as less stable long-term. This procedure is very risky and can therefore not be recommended.

8.4 Time Temperature Superposition (TTS)

Dynamic mechanical measurements of polymers at different temperatures and frequencies give the same moduli curves but they are shifted on the time scale. This means that polymers exhibit a material-specific behavior that is independent of the prevailing temperature. In other words, the rheological curve does not change but is shifted along the x-axis in accordance with the prevailing temperature. Consequently, the time scale can be expanded considerably by varying the temperature [81] and subsequently shifting the measuring curve relative to a reference temperature T_{ref}. In this way a curve is obtained relative to a reference temperature that might be impossible to measure with normal measuring times. This type of data plot is referred to as a master curve.

As always, the validity of a theory like this depends on several important boundary conditions that must be kept constant. The sample may not enter into any chemical or physical reaction during the measurement. TTS theory is applicable only in the viscoelastic range, i.e. in the plateau region. Emulsions also exhibit this kind of plateau (see Sect. 8.3). This theory was originally formulated by Williams, Landel and Ferry [82].

8.4.1 Softening Point

To check the applicability of the TTS principle [83], a solid w/o emulsion was chosen that had been stored for one year under defined conditions and met the legal shelf-life requirement of > 36 months. To exclude structural changes due to temperature effects, first the softening and freezing points were determined. The softening point was determined (Fig. 8.41) in a temperature test with a starting temperature of $T_{\text{Start}} = 25\,°\text{C}$, a heating rate of $\Delta T = 1\,°\text{C}$ and a final temperature of $T_{\text{end}} = 80\,°\text{C}$.

Two softening points were found, T_{soft1} at ca. 27 °C and T_{soft2} at ca. 47 °C.

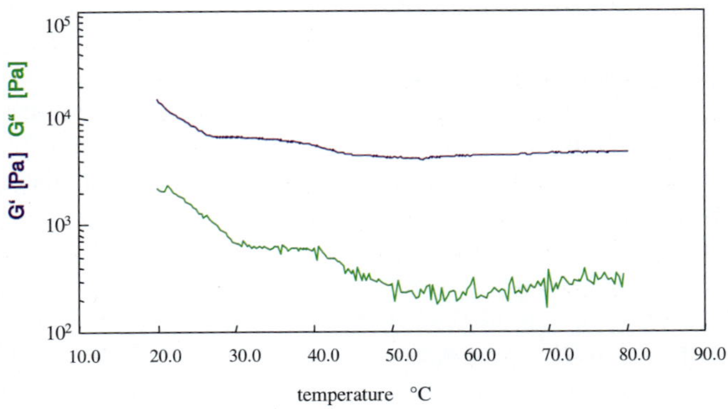

Fig. 8.41. Softening point

8.4.2 Freezing Point

A temperature test was also used to determine the freezing behavior. It was also performed with a starting temperature of $T_{\text{Start}} = 25\,°\text{C}$ but the cooling rate was $\Delta T = -1\,°\text{C}$ and the final temperature $T_{\text{end}} = -50\,°\text{C}$.

The freezing point (Fig. 8.42) was $T_{\text{fr-}} = -30\,°\text{C}$. With these two preliminary experiments the temperature range was determined in which no chemical changes should be expected.

8.4.3 Determination of the Master Curve at Constant Frequency

The time-dependent measurements were performed in the temperature range $T = -10$ to $55\,°\text{C}$. A description of the four methods for determination of the shift factors follows.

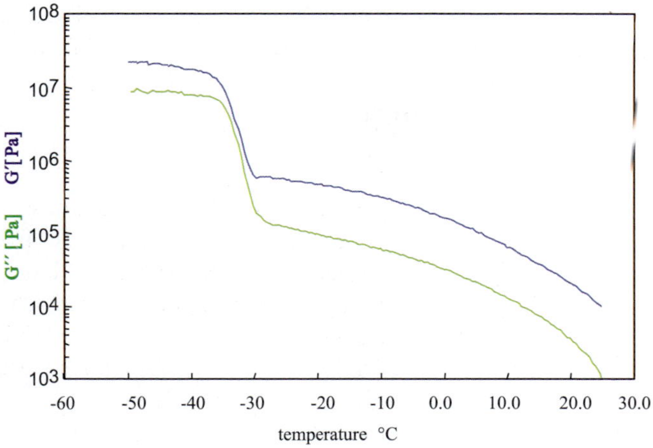

Fig. 8.42. Freezing point

8.4.3.1 Determination of the Activation Energy via the Temperature

Many chemical reactions do not proceed spontaneously even if there is a high potential that heat of reaction could be released. The input energy causes the chemical systems present to rearrange, forming an intermediate state in the process. The activation energy is the amount of energy needed to initiate a chemical reaction. The heat of reaction released during steps of a reaction serves as a continuous source of energy, ensuring the overall chemical reaction goes spontaneously to completion.

The following equation is known from the literature:

$$a_T = \left(\frac{1}{T_2} - \frac{1}{T_1}\right) = \frac{\ln a_T \cdot R}{E} \tag{8.6}$$

where T_1 and T_2 are not defined more specifically. For our calculations the next respective T_1 value was always used as T_2. This gave the following values:

The resulting shift factors are very small and consequently the master curve calculated accordingly corresponds to a shelf-life of the emulsion of 10^2 min, which is about 1.5 h.

If we were to define T_2 instead as the reference temperature of 20 °C, this would not change in the values of the shift factors substantially.

8.4.3.2 Viscosity

Viscosity function curves measured at different temperatures can be shifted approximately to a master curve in a plot using the shear rate:

$$a_T \equiv \frac{\eta_0(T)}{\eta_0\left(T_0\right)} \tag{8.7}$$

Table 8.9. Determination of the shift factors by means of the temperature

T [°C]	$1/T_1$	$1/T_2$	$(1/T_1 - 1/T_2)$	$(1/T_1 - 1/T_{0,20})$
−10	−0.1000	−0.2000	0.1000	−0.1500
−5	−0.2000	–	–	−0.2500
0	–	0.2000	–	–
5	**0.2000**	**0.1000**	**0.1000**	**0.1500**
10	**0.1000**	**0.0667**	**0.0333**	**0.0500**
15	**0.0667**	**0.0500**	**0.0167**	**0.0167**
20	**0.0500**	**0.0400**	**0.0100**	**0.0000**
25	**0.0400**	**0.0333**	**0.0067**	**−0.0100**
30	**0.0333**	**0.0286**	**0.0048**	**−0.0167**
35	**0.0286**	**0.0250**	**0.0036**	**−0.0214**
40	**0.0250**	**0.0222**	**0.0028**	**−0.0250**
45	**0.0222**	**0.0200**	**0.0022**	**−0.0278**
50	**0.0200**	**0.0182**	**0.0018**	**−0.0300**
55	0.0182			−0.0318

The complex dynamic viscosity is obtained from oscillatory measurements. The complex dynamic viscosity $\eta^*(\omega)$ corresponds, however to the viscosity η measured at steady shear rates if the Cox/Merz relation is valid ($\dot{\gamma}$).

The Cox/Merz relation [84] is not valid for most emulsions! This rule is only valid for other polymer fluids [85]:

$$|\eta^*| = \eta \quad \text{für} \quad \omega = \dot{\gamma} \tag{8.8}$$

If the Cox/Merz relation is valid, the following is true:

$$a_T \equiv \frac{\eta_0(T)}{\eta_0(T_0)} = \frac{\eta_0^*(T)}{\eta_0^*(T_0)} \tag{8.9}$$

The reference temperature is 20 °C.

The shift factors determined are too small to be able to predict a shelf-life < 100 min.

8.4.3.3 Arrhenius Equation

The ability of atoms or lattice defects to move in a crystal depends on the ratio of the activation energy [82] needed to the thermal energy present.

In an Arrhenius plot the logarithm of a variable is plotted vs. the reciprocal temperature $\left(1/T\left[\mathrm{K}^{-1}\right]\right)$.

In the plot against $1/T$, y_0 is the y-intercept and E_0/R the slope of the straight line. E_0 is the activation energy (or free energy of formation) of the process. The Arrhenius plot is one of the most important analytical tools in many fields, from physics to biology. The temperature is always stated in Kelvin.

Table 8.10. Determination of shift factors using the Cox/Merz relation

T [°C]	η^* [Pa]	a_T
−10	232 000	15.49
−5	238 900	15.95
0	178 800	11.94
5	91 960	6.14
10	51 830	3.46
15	26 620	1.78
20	14 980	1.00
25	8974	0.60
30	7146	0.48
35	6330	0.42
40	5071	0.34
45	4164	0.28
50	3461	0.23
55	3423	0.23

$$\eta_0(T) = \eta_0\left(T_0\right) \cdot e^{\left[\frac{E_0}{R}\left(\frac{1}{T}-\frac{1}{T_0}\right)\right]} \tag{8.10}$$

From the viscosity equation, the following is derived:

$$\ln a_T = \frac{E_0}{R}\left(\frac{1}{T} - \frac{1}{T_0}\right) \tag{8.11}$$

Exception: For a fully cross-linked polymer it is possible to calculate the apparent viscosity from the results of an oscillatory measurement:

$$\left|\eta^*\right| = \frac{|G^*|}{\omega} = \frac{\sqrt{G'^2 + G''^2}}{\omega} \tag{8.12}$$

In a fully cross-linked polymer the elastic portion is several times greater than the viscous portion. The viscous portion is negligible. The measured values are recorded at a frequency of $\omega = 1$ rad/s:

$$G' \gg G'' , \quad \omega = 1\,\text{rad/s} \tag{8.13}$$

$$\left|\eta^*\right| = \left|G'\right| \tag{8.14}$$

$$\ln \frac{\eta_0^*(T)}{\eta_0 * \left(T_0\right)} = \frac{\ln G'}{\ln G'_0} = \ln G' - \ln G'_0 = \ln G'_1 \tag{8.15}$$

$$\ln G'_1 = \ln a_T = \frac{E_0}{R}\left(\frac{1}{T} - \frac{1}{T_0}\right) \tag{8.16}$$

$$a_T = 10^{\ln G'_1} \tag{8.17}$$

The shift factors are still always too small to significantly change in the shelf-life.

Table 8.11. Determination of the shift factors using the Arrhenius equation

T [°C]	G' [Pa]	$\ln G'$	$\ln G'_1$	a_T
−10	231 100	12.35	2.74	550.31
−5	238 200	12.38	2.77	590.02
0	178 300	12.09	2.48	302.85
5	91 290	11.42	1.81	64.83
10	51 300	10.85	1.24	18.20
15	26 400	10.18	0.57	3.72
20	14 890	9.61	0.00	1.00
25	8948.5	9.10	−0.51	0.31
30	7137.1	8.87	−0.74	0.18
35	6324	8.75	−0.86	0.14
40	5067.7	8.53	−1.08	0.08
45	4162.8	8.33	−1.28	0.05
50	3460.7	8.15	−1.46	0.03
55	3422.9	8.14	−1.47	0.03

8.4.3.4 WLF Equation

The temperature dependence of the shift factors can be described by the semi-empirical equation of Williams, Landel and Ferry (WLF equation) [82] or the Vogel–Fulcher equation:

$$a_T = e^{\left[\frac{-c_1(T-T_0/}{c_2+(T-T_0/}\right]} \tag{8.18}$$

$$\ln a_T = \frac{-c_1\left(T-T_0\right)}{c_2+\left(T-T_0\right)} \tag{8.19}$$

The material-specific constants can be determined by plotting $T-T_0/\ln(a_T)$ against $T-T_0$. Since a_T is not a known value but one that also needs to be determined, this equation cannot be used to calculate shift factors.

8.4.3.5 First Conclusion

We have examined four methods [83] for determination of shift factors and not one of them proved to be suitable. The calculated shift factors were so low with all methods that the shelf-life determined was much shorter than the known shelf-life of the emulsion. A possible explanation for the considerable deviations of the calculated values from the actual values could be that the emulsion had been stressed so much by the forces introduced during filling that the 100 min calculated give the time elapsing until the emulsion has reached equilibrium again. To determine this delay time (waiting time before the start of the test), a creep or creep-recovery test was performed. The test gave a waiting period of 8 min. Therefore this can be eliminated as the cause.

8.4.3.6 Determination of the Master Curve with Variable Frequency

In the second test 13 different frequency scans were performed at constant temperature. This technique makes it possible to transfer polymer data to a frequency range that is not accessible in a real measurement. Since the strain depends on the temperature, the measurement is divided into two temperature ranges.

Temperature range −30 to +25°C

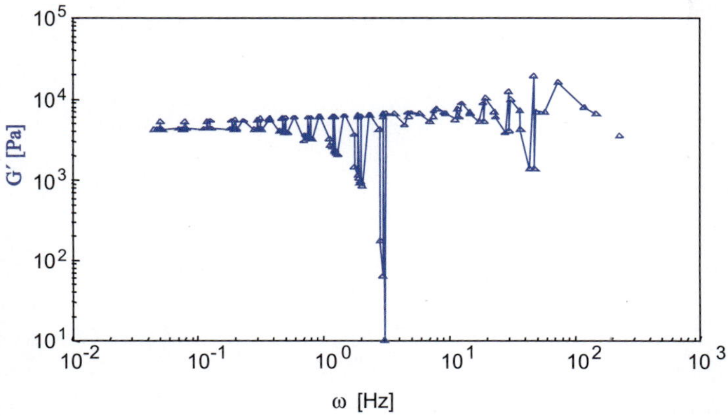

Fig. 8.43. Master curve from frequency-temperature measurements

Temperature range −25 to +100°C

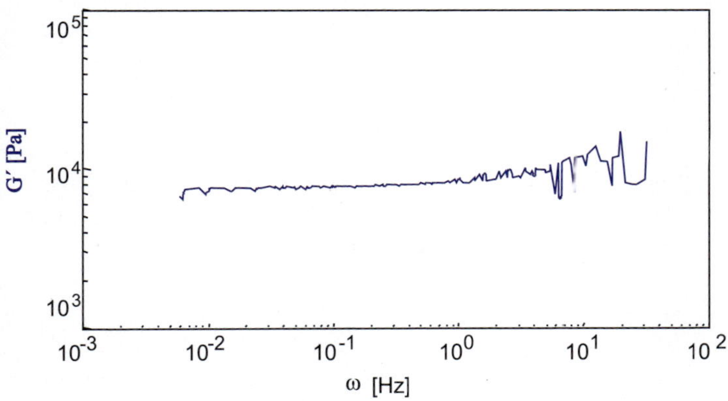

Fig. 8.44. Master curve from frequency-temperature measurements

In the temperature range from −30 to +25 °C the master curve is not linear. At T = +25 to +100 °C it is linear in the range from 10^{-2} to 10^{0}. This would give a shelf-life of ca. 630 s, i.e. 10.5 min. Since the solid sample was used also in this test, TTS cannot be applied here either.

8.4.3.7 Final Conclusion

The time temperature superposition principle provides no practice-relevant results for the sample studied. Since the w/o emulsion is a very firm cream known to have a long shelf-life, it can be concluded that the TTS principle is not generally applicable for emulsions.

9 Interpretation

As already pointed out several times, rheological measurements are extremely sensitive to changes of every type. For instance, different results can arise from use of a raw material from another vendor. Even just a slight change in the production process can lead to dramatic changes in the physical properties. It is just this sensitive behavior that gives rise to the only – and serious – disadvantage of rheology. A rheological measurement detects the effect but not the cause. This applies not just to cosmetic emulsions. Nevertheless some general information can be obtained from certain relationships.

9.1 Relationships for Polymers

For example, the following relationships are known from the polymer area:

1. The solids content of a polymer solution depends on the zero shear viscosity. With decreasing solids content the zero shear viscosity decreases and the Newtonian region becomes larger at low concentrations because the Newtonian character of the solvent increasingly dominates with decreasing solids content.
2. The molecular weight depends on the zero shear viscosity. With increasing molecular weight the zero shear viscosity also increases. A curve similar to that in Fig. 9.1 is obtained but with other viscosity values. Another method often used to determine the molecular weight from the viscosity data is to measure a dilution series and extrapolate the results to zero concentration [86, 87] in a so-called Huggins plot (Fig. 9.2). In Fig. 9.2, three different types of rubber with a molecular weight difference of 200 000, 250 000 and 500 000 are presented.
3. The molecular weight dependency is a function of the frequency of dynamic crossover ($G' = G''$). With increasing molecular weight the frequency of dynamic crossover is shifted to lower frequencies.
4. The molecular weight distribution [88] depends on the absolute value of dynamic crossover. The narrower the molecular weight distribution (decreasing dispersity), the higher is the absolute value of crossover.
5. The increase in the moduli provides information on the degree of crosslinking.

For an uncrosslinked polymer solution the viscous part (top curve in Fig. 9.4) is larger than the elastic part over the entire frequency range. Therefore the loss

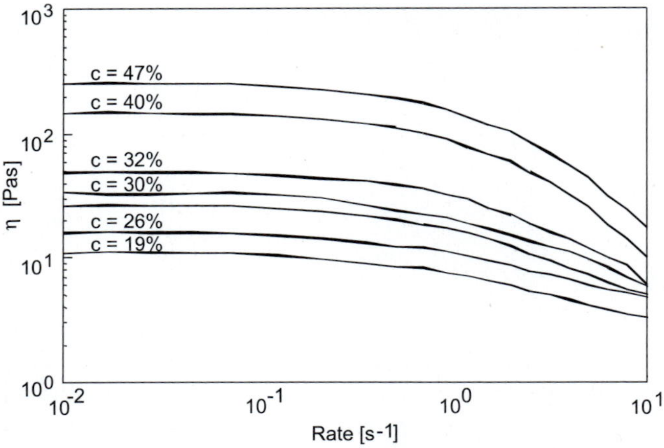

Fig. 9.1. Polymer solution as a function of solids content

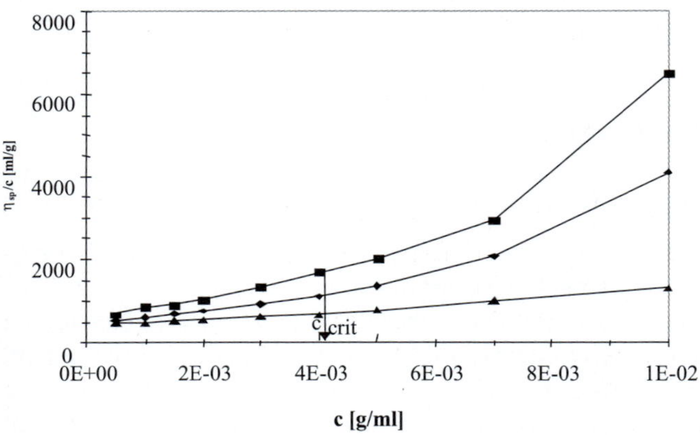

Fig. 9.2. Determination of the molecular weight from a dilution series

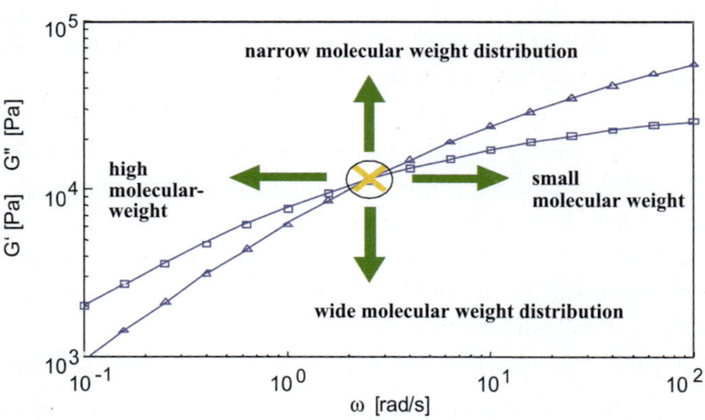

Fig. 9.3. Dynamic mechanical molecular weight dependency

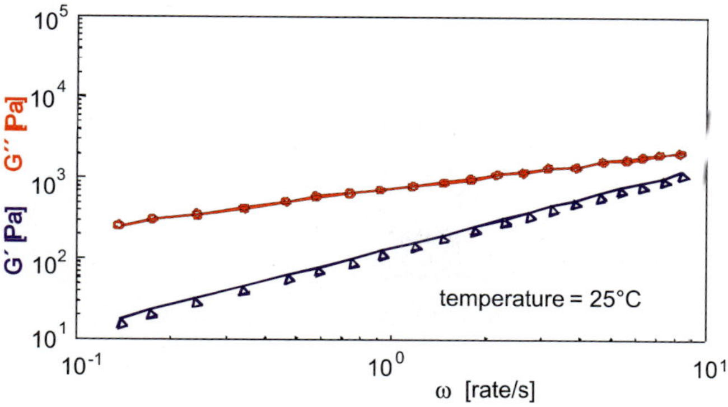

Fig. 9.4. Uncrosslinked polymer solution

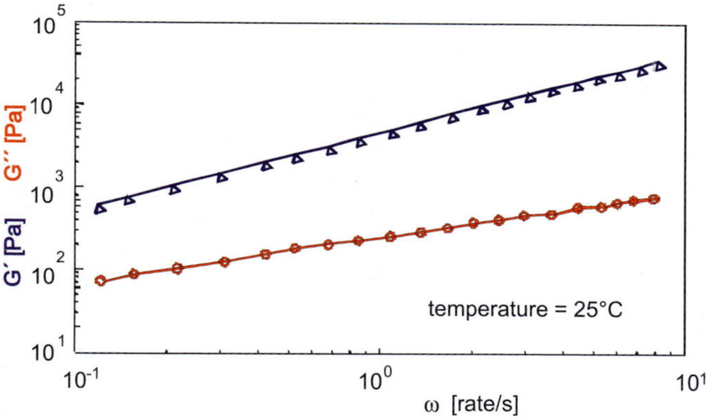

Fig. 9.5. Partially crosslinked polymer

modulus G'' is $\gg G'$ because, as the term polymer solution [89] already implies, it is mainly a liquid (a small portion of a polymer was placed in a solvent).

In the case of a partially crosslinked polymer the storage modulus G' is larger than the loss modulus G'' in the frequency range measured, and the slope of the storage modulus curve G' is greater than that of the loss modulus G''.

In the case of a fully crosslinked polymer the moduli are very large and the curves for the storage and loss moduli run nearly parallel, with a difference of more than one power of ten between the absolute values. This is a solid.

9.2 General Statements for Cosmetic Emulsions

1. The yield stress is especially well suited for classification of the widest variety of emulsion types. Different emulsion classes can be recognized quickly and easily by plotting the viscosity as a function of shear stress. For instance the creams in Fig. 9.7 show a high viscosity at high shear stress whereas the milk

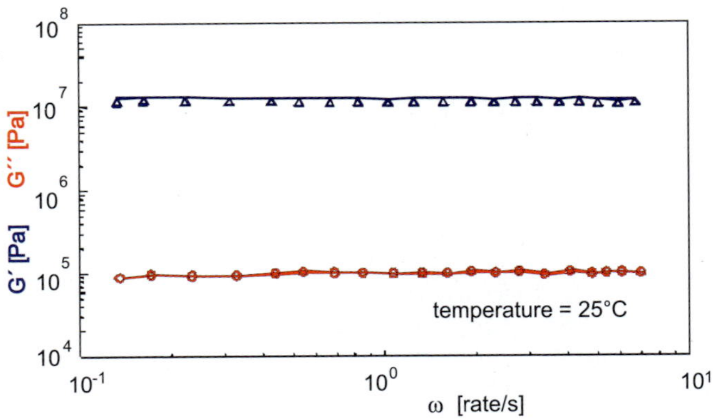

Fig. 9.6. Fully crosslinked polymer

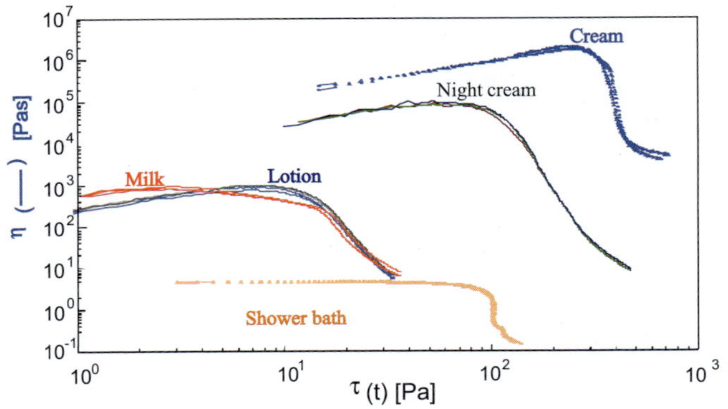

Fig. 9.7. Yield stress for classification of products

and lotion are found at a low viscosity and low shear stress, as a customer would expect of the respective products.

2. As known from the polymer area, only limited information on the structure or gel network of an emulsion can be obtained from viscosity curves. In Figs. 9.8 and 9.9 each of the viscosity curves is presented, with the non-linear behavior recognizable.

In contrast, dynamic mechanical measurements allow determination of characteristic properties of different products. Presented in Fig. 9.8 is a typical curve showing the frequency dependence of an o/w emulsion. At low frequencies, meaning low stress, the two moduli are nearly equal. The properties of the storage modulus do not dominate until higher frequencies and do this despite the fact that the viscosity curve decreases continuously.

A very solid gel network can be seen in Fig. 9.9, for the storage modulus G' is clearly larger than the loss modulus G'' over the entire frequency range measured, and the two curves run nearly parallel over a wide range of frequencies. Moduli

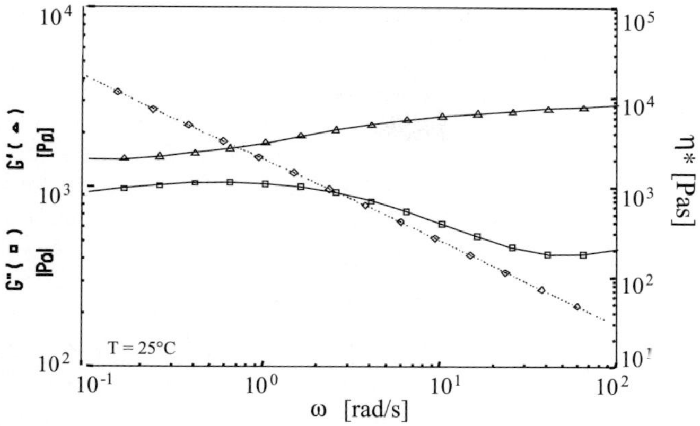

Fig. 9.8. Typical storage and loss modulus curves for an o/w emulsion

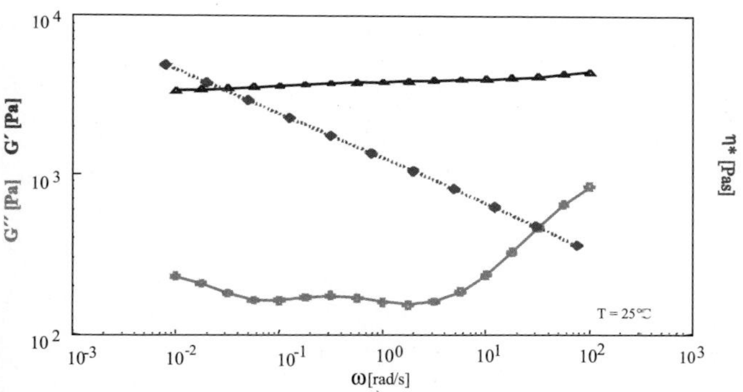

Fig. 9.9. Typical storage and loss modulus curves for a w/o emulsion

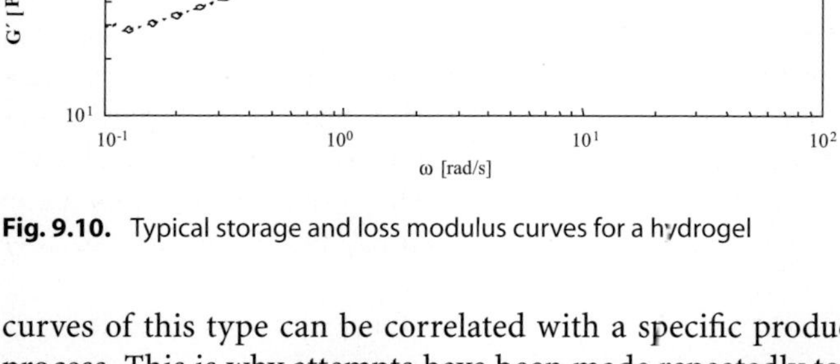

Fig. 9.10. Typical storage and loss modulus curves for a hydrogel

curves of this type can be correlated with a specific product and its production process. This is why attempts have been made repeatedly to obtain this data with fast, simple measurements during the production process.

In Fig. 9.10 the curve of a hydrogel is shown. Crossover, i.e. the intersection of the two moduli, is typical of this product. At lower frequencies the loss modulus G'' dominates, but from a certain frequency the storage modulus G' then dominates. In this case it is actually very easy to derive a method that allows measurement of informative results during the production process easily and quickly.

As evident from preceding figures, it has not always been possible to find these characteristic properties and then adapt them accordingly to the production process. Searching for such characteristic properties is the rheologist's job.

10 Calibration/Validation

Calibration is establishment of the relation of an input to an output variable or of the indicator of a measuring instrument to the measured value itself. For the scales named, calibration establishes the indicator error of a measuring instrument.

Validation is establishment of the accuracy of a scientific experiment or measurement method. Validation is based on agreement of a test result with a criterion independent of the actual test series such as the accuracy of prediction obtained by another pathway. A standard is used for this. One example of such a standard is the standard meter.

The precursor of the standard meter was the prototype meter chosen by the first General Conference on Weights and Measures on 26 September 1889. The standard meter is a rod of hammered platinum sponge with a final length of 1 m (rectangular cross-section 25.3 × 4.0 mm). It should be 1/40 millionth of an earth meridian. The prototype meter is a rod constructed as a measuring stick with an S-shaped cross-section (20 × 20 mm) made of an alloy of 90% platinum and 10% iridium. It was supposed to reproduce the standard meter [90] as accurately as possible and was both the definition of and standard for the unit of length until 1960. This standard meter was kept in the Bureau et Mesures in the Pavillion de Breteuil de Poids in Sèvres in Paris at 0 °C.

Since 14 October 1960 a new definition of the meter has been valid in accordance with the decision of the 11th General Conference of Weights and Measures. This is based on the wavelength of the radiation emitted by the krypton 86 atom:

$$1\,\mathrm{m} = 1\,650\,763.73\,\lambda_0$$

where λ_0 is the wavelength in a vacuum of the radiation between the energy states $5d_5$ and $2p_{10}$ of the krypton 86 atom, which is the orange line at 605.6 nm.

As the history of the standard meter shows, standards can also change. This is why it is important to state what the measuring result was compared with.

For the field of rheology there are several companies and institutes offering viscosity standards including:

NIST = National Institute of Standards and Technology
PSS = Polymer Standard Services
PTB = Physikalisch-Technische Bundesanstalt Braunschweig
Brookfield Viscosity Standard
Cannon Viscosity Standard

In principle, it is possible to buy standards from every rheometer manufacturer or to ask them where such standards can be obtained. Differences exist especially in price but also in quality.

Unfortunately, it cannot be assumed that every viscosity standard is a Newtonian fluid. Therefore we can legitimately ask about the valid range (shear rate range). It is not always necessary to buy the most expensive standard, but it is important that a Newtonian standard is accompanied by a certificate stating the dynamic viscosity η at different temperatures and possibly also the density ϱ as well as the kinematical viscosity ν of the standard.

The dilemma we need to solve is which viscosity standard to buy or in other words which viscosity our standard should have. This will depend on the products we want to measure with our viscometer. Make a list of the products with their estimated viscosities. In Table 10.1 some viscosity [91] data is given for typical products.

Table 10.1. Typical viscosity data at $T = 20\,°\text{C}$

Petroleum	0.65 mPas
Water	1.00 mPas
Mercury	1.5 mPas
Grape juice	2 to 5 mPas
Blood (37 °C)	4 to 15 mPas
Cream	10 mPas
Olive oil	100 mPas
Honey	10 000 mPas
Tar	1 000 000 mPas
Bitumen	10 000 000 mPas

From this list choose a viscosity standard (taking into account the respective measuring temperature) from the lower, middle and highest viscosity range. Before ordering the three viscosity standards you should also think about quantities because it is not the instrument that will be checked but the instrument together with a measuring system. In Sect. 5.3–5.6 we were introduced to the three most important systems, namely the cone-plate, parallel plate and cylinder systems. Each of these systems comes in very different geometries. For instance parallel plate systems have dimensions ranging from 4 to 60 mm. Therefore we need to estimate the sample amounts to be measured and think about how often we want to perform the measurements.

We also need to think about the test procedure or how we want to design comparative measurements. To do this we need to make a test plan or in this case a validation plan. The validation plan in Table 10.2 contains, in addition to the three viscosities we want to measure, possible temperature variants (in this case $T = 20$, 25 and 40 °C) as well as information about the measuring

Table 10.2. Possible validation plan

System A

	Low viscosity 20 °C	Low viscosity 25 °C	Low viscosity 40 °C	Middle viscosity 20 °C	Middle viscosity 25 °C	Kiddle viscosity 40 °C	High viscosity 20 °C	High viscosity 25 °C	High viscosity 40 °C
PP									
CP									
ZS									

System B

	Low viscosity 20 °C	Low viscosity 25 °C	Low viscosity 40 °C	Middle viscosity 20 °C	Middle viscosity 25 °C	Middle viscosity 40 °C	High viscosity 20 °C	High viscosity 25 °C	High viscosity 40 °C
PP									
CP									
ZS									

System C

	Low viscosity 20 °C	Low viscosity 25 °C	Low viscosity 40 °C	Middle viscosity 20 °C	Middle viscosity 25 °C	Middle viscosity 40 °C	High viscosity 20 °C	High viscosity 25 °C	High viscosity 40 °C
PP									
CP									
ZS									

system and instrument. The only thing we need to do now is determine how often the measurements should be repeated. We can obtain this information from statistics, which is why we will briefly consider some basic statistical principles here.

10.1 Basic Principles of Statistical Analysis

We will now examine the most important terms such as the mean, true value, standard deviation, coefficient of variation, variance, measured value, population, sample size, range and data number.

10.1.1 Normal Distribution (Gaussian Distribution)

Normal distribution is a distribution function of a random variable [92] where the density function takes the shape of a Gaussian bell curve (normal distribution curve). A variable of normal distribution x with the expected value μ and variance σ^2 has the density function:

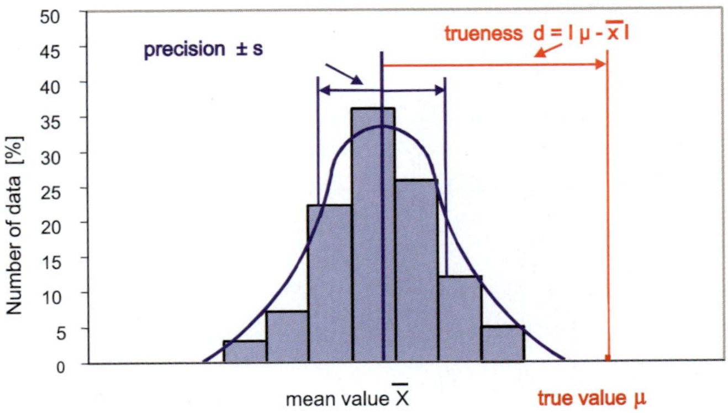

Fig. 10.1. Overview of several statistical terms

$$f(x) = \frac{1}{\sigma\sqrt{2\pi}} \mathrm{e}^{-\frac{(x-\mu)^2}{2\sigma^2}} \tag{10.1}$$

where e is the base of the natural logarithm. It is symmetrical around the axis $x = \mu$, reaches its maximum there and has two inflection points (at $x = \mu \pm \sigma$). For $\mu = 0$ and $\sigma = 1$ a standardized normal distribution is obtained. In traditional physics it is assumed that many phenomena in nature can be meaningfully described by a normal distribution. Many distributions in statistical practice are normal, or at least nearly normal, and therefore the distribution of the measured data (for $n > 30$) must be nearly normal.

10.1.2 Mean Value

The mean is also often referred to as the average or arithmetic mean. It is calculated by summing all data and dividing this by the data number. For n data x_i the following equation results:

$$(\mu) = \overline{\chi} = \widehat{\chi} = \frac{\sum_{i=1}^{n} \chi_i}{n} \tag{10.2}$$

Fig. 10.2. Numbers thrown for 12 dice

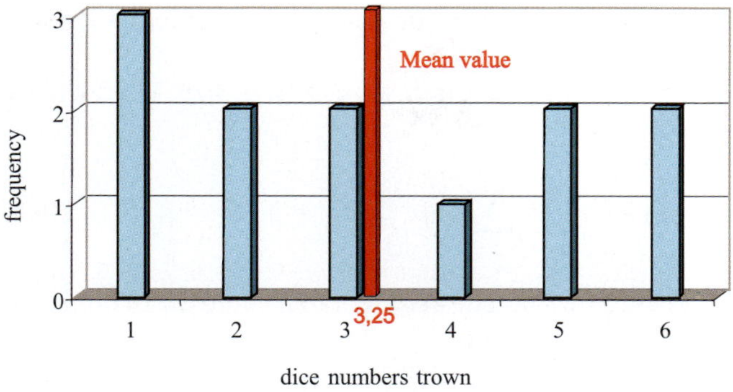

Fig. 10.3. Frequency of the numbers thrown and their mean

As an example let us consider a dice experiment [93] in which the following numbers are thrown consecutively:

The mean of the dice experiment (μ) in Fig. 10.2 is

$$(\mu) = \overline{\chi} = \widehat{\chi} = \frac{39}{12} = 3.25 \tag{10.3}$$

10.1.3 True Value

- The true value is a purely theoretical number and rarely known exactly. It is the value we would obtain in a perfect measurement. True values are by nature indefinite.
- Example 1: In our dice game the true value is 3.5. This means that we should obtain a mean of 3.5 for an ideal dice if we through the dice an infinite number of times. If we do not obtain 3.5 as the mean after an infinite number of throws, the difference is the so-called bias.

10.1.4 Standard Deviation and Variance

Another characteristic parameter besides the mean is the degree of fluctuation (standard deviation or variance). Variance is a measure of how the individual data are distributed around the mean (how strongly the data fluctuate around the mean).

The following is valid for the variance: var(χ) from n data : χ_i

$$\text{var}(\chi) = \frac{1}{n-1}\sum_{i=1}^{n}(\chi_i - \overline{\chi})^2 = \frac{1}{n-1}\left[\sum_{i=1}^{n}\chi_i^2 - n\chi^2\right] \tag{10.4}$$

We divide here by $n-1$ because the mean used in the equations was calculated from the data. This reduces the number of degrees of freedom. For calculation of the variance the square is used because then the points lying farther away more strongly influence the result.

In our dice experiment the calculated mean we used was 3.25. The variance derived from this is therefore

$$s^2 = var(\chi) = \frac{1}{12-1}\sum_{i=1}^{n}(\chi_i - 3.25)^2 = 3.66 \qquad (10.5)$$

If we had based our calculation on the true value μ rather than on the mean $\overline{\chi}$, then we would not divide by $n - 1$ but by n. We often speak of the error of each measurement or the true value.

Here we use Greek symbols (reference to the true value) rather than Roman letters (reference to the calculated mean): σ^2 (variance) or σ (standard deviation):

$$\sigma^2 = \frac{1}{n}\sum_{i=1}^{n}(\chi_i - \mu)^2 = \frac{1}{n}\left[\sum_{i=1}^{n}\chi_i^2 - n\mu^2\right] \qquad (10.6)$$

The true value in our dice game is characterized by assuming an ideal dice and obtaining a mean of 3.5 with many throws. The true value is then 3.5.

10.1.4.1 Standard Deviation

In practice, the standard deviation [94] is used more often than the variance because it has the same dimension as the measured values.

The variance is a measure of how the individual data are distributed around the mean.

The standard deviation $s(x)$ or $sdv(x)$ is the square root of the variance and is often referred to as the mean square error of the individual data:

$$s(\chi) = sdv(\chi) = \sqrt{var(\chi)} = \sqrt{s^2(\chi)} \qquad (10.7)$$

This value does not depend on the sample size but is influenced by the quality of a measurement method. We can see from the equation that an increasing number of measured values is compensated by error summation.

10.1.4.2 Coefficient of Variation

The coefficient of variation is a measure of the scatter and gives the difference between the smallest and largest value (also called range).

10.1.5 Measured Value, Result, Random Variable

Measured value: measured, observed or read values. It is a quantity that is measured.

Result: result of an analysis after doing a measurement and all subsequent evaluation steps.

Variate (random variable): numerical value of a measured value or of a result. Characteristic of a concrete quantity that differs from test object to test object.

In particular, these can:

- Be very specific discrete values (dice game, indicator of a digital clock);
- Take on any intermediate value continuously within a defined, limited range (pointer of an analogue clock); or
- Be variable (for every value) around this specific value with the probability decreasing continuously with increasing distance, but theoretically cannot be the value 0 (reading of an analogue measuring instrument).

10.1.6 Population, Series, Measured Value

Every variate belongs to an infinite number of possible variates. This is called the population. A limited part of this population is called a series. The occurrence of a variate can be described by a specific probability function that describes the distribution of the variates around an expected value for the population (calculated value, as close to the true value as possible).

We use terms like mean to characterize discrete distributions, whereas an expected value can describe both discrete and continuous distributions.

10.1.7 Errors and Deviations

The expected value or mean differs from the true value by an error. In contrast, a deviation is the difference between the variate and the mean within a series. In the dice game this would be the deviation from 3.5 for every value thrown. The systematic error (bias) is the difference between the mean and the true value. In our dice game example the bias would be 3.5 – 3.25 = 0.25.

Table 10.3. Overview of different terms used for error

Error	$\delta = x - \mu_x$
Bias	$b = \overline{x} - \mu_x$
Deviation	$d_i = x_i - \overline{x}$

10.1.7.1 Error Types

Besides gross errors (using the wrong measuring stick or a tape measure for micromeasurements), two types of errors [95] are important:

- Statistical (random) errors:
  These determine the precision (reproducibility) of a method; they usually cannot be completely avoided and often cannot be characterized. The precision is a measure of the scatter among variates.

- Systematic errors (bias):
  These influence the accuracy or trueness of an analytical method; they are deviations of the results from the true value and caused by interfering factors (unexpected components) or incorrect measuring techniques (wrong method, defective instrument, or bent measuring stick). They can only be discovered by comparative measurements with another technique or in another laboratory with round robin experiments, but not by repeated measurements. The mean is often used to represent the expected value of a test result. In this case the bias represents the deviation of the mean of a number of results from the established standard value. In analytical chemistry this is usually called the systematic error.

Often the accuracy is also used as the total error. Unfortunately, the DIN and ISO standards differ in this point.

To illustrate both types of errors the pattern on a target is frequently used:

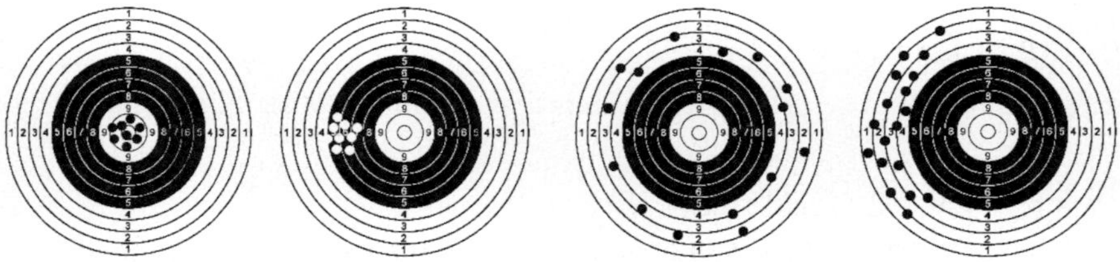

Fig. 10.4. Precise and true; precise and false; imprecise and true; imprecise and false

Since the true value usually is not known, an attempt is made to approximate this value as closely as possible by reducing all sources of bias and reducing the uncertainty of the measured value by doing many measurements.

10.1.8 Precision

- Precision is a measure of the agreement among test results measured independently under fixed conditions.
- Precision can be stated for every data set obtained under defined but freely chosen conditions. The precision of a set of results can be quantified as the standard deviation.
- Exact results for an analyte are characterized by a small statistical variation. This can be indicated by a small standard deviation, a narrow range of values or a small difference between quartiles. Each is a measure of precision. Precision is a characteristic of a data set.
- The calculation of precision should be based on at least *eight* independent measurements.

10.1.9 Accuracy

- Accuracy is a measure of the agreement of the (individual) test result and the true value of the measurable variable.
- Accuracy is a parameter that combines precision and trueness (i.e. the effects of random and systematic error).

Assuming the results of a specific measurement has a bias of zero or a very small bias (i.e. it is the true value), the accuracy corresponds to the precision. However, if the precision is poor, the individual results will be inaccurate and deviate greatly from the conventional value. If the precision is good, the result will be accurate if there is no bias. However, if the method has a large bias, even the results with a good (small) precision will be inaccurate.

By convention, the accuracy is reported relative to a measured value. A result of 10 µg/L with an accuracy of 1 µg/L accordingly has an accuracy of 10%. (Although this numbers game describes an inaccuracy, we speak of accuracy.)

10.1.10 Trueness

- Trueness is a measure of the agreement between the mean obtained from a large data set and the accepted reference value. (The accepted reference value is equated with the conventional value.)
- A true result is a result with a small (or ideally no) systematic error regardless of the distribution of the results. Perfect trueness cannot be achieved. Therefore, in terms of analysis trueness is always trueness within defined limits. These limits can be broader for high concentrations than for trace amounts. It is also important to distinguish between trueness and accuracy. Accuracy is a measure of the agreement between a single test result and the accepted reference value (true value).
- Example:
  Let us assume we obtain the following results [93] (in arbitrary units) from a set of ten independent measurements:
  25.35 27.35 24.89 25.23 26.48 26.08 25.98 27.82 25.55 26.33.
  The mean is 26.11. If the reference value is 26.95, then the trueness is: 26.95 – 26.11 = 0.84.
  The accuracy of the second measurement would be: 26.95 – 25.58 = 1.37, etc.

10.1.11 Repeatability

- Repeatability is the precision under repeatable conditions.
- Repeatability is a measure of the agreement among the results of independent measurements of the same analyte when all of the following conditions are met:
  - Use of the same test method,
  - By the same person,

- With the same measuring instrument,
- In the same location,
- Under the same test conditions, and
- Repeated within a short time interval.

Independent measurements are performed on different sub-samples of a test material. If possible, at least eight measurements should be performed. The repeatability is a characteristic of a method and not of a result.

10.1.12 Reproducibility

- The reproducibility is the precision under reproducible conditions.
- The reproducibility is a measure of the agreement among the results of independent measurements of the same analytes in sub-samples of a test series, where the individual measurement are subject to different conditions such as other analysts, other measuring instruments, other locations, other test conditions or other times; the same method is always used.

Methods that have a high reproducibility cannot be used for a valid comparison in a real situation. In this case either the method must be improved or another method with a lower reproducibility must be used.

10.1.13 Outliers

- Outliers are measured values that differ significantly from other measured values and are identified by outlier tests.
- There are many outlier tests (e.g. Nalimov), but all are problematic. Either the assumed distribution is false or a systematic error is present.
- Some experimenters permit no outliers, contending one must check if there is a real, meaning systematic, error present. If this is the case, then the data point must be removed from the data set. If no bias can be detected the data point may not be removed. Most outlier tests are based on distributions and definitions of confidence intervals.

10.2 Back to the Laboratory

Why is calibration or validation so important? There are many instrument manufacturers and just as many measurement methods and test instructions specifying how and what we should measure to solve a certain problem. Nothing is specifically said about the precision and accuracy but they are tacitly assumed. Therefore it is important to know how accurately a measurement can be performed and the error associated with each measuring instrument in order to be able to compare results from different measuring instruments and systems. This is also why it is necessary to invest the time and effort needed for calibration and validation, at least when a new measuring instrument is put into operation.

We use statistics for this as a mathematical tool. Based on the statistics we can now complete the validation plan. We have learned that the number of measurements for the first validation should be $n > 30$.

For subsequent re-validations this number can be reduced to $n = 5$. If large deviations occur, repetition of the full validation process will be unavoidable.

In the viscosity curves shown in Fig. 10.5 the Newtonian character of the oils is evident at least at $T = 25\,°\text{C}$. In addition, we can assume that the reproducibility is good. However, in Fig. 10.6 we will now look more closely at one result.

According to the certificate of analysis the nominal viscosity at $T = 25\,°\text{C}$ is $\eta = 985.05\,\text{mPas}$. The measured viscosity is $\eta = 925.80\,\text{mPas}$ with a standard deviation of $\sigma = 2.72\%$.

The distribution of the measured data is nearly normal and the reproducibility fairly low. However, the viscosity we measured with this instrument is too low,

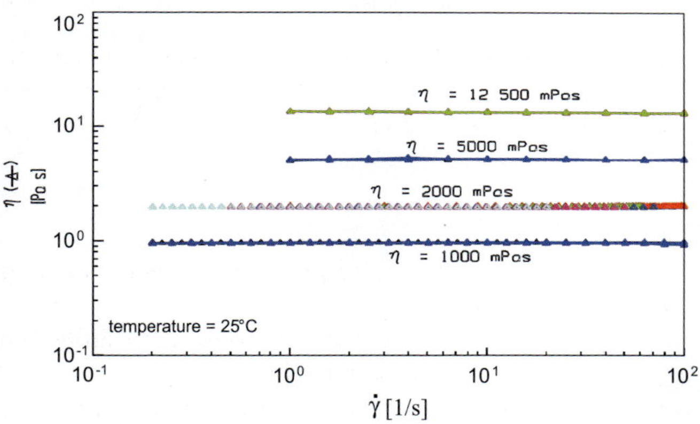

Fig. 10.5. Validation with different calibration oils with $n = 35$ measurements for each

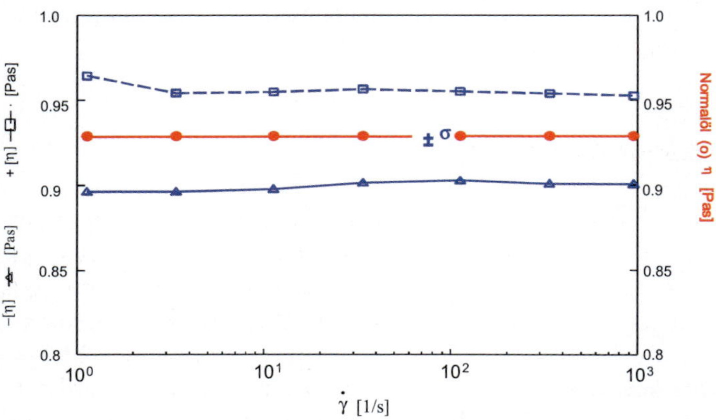

Fig. 10.6. Mean and standard deviation of the standard oil with a nominal $\eta = 985\,\text{mPas}$

namely: (nominal viscosity η = 985.05 mPas) – (mean viscosity η = 925.80 mPas) = 59.7 mPas.

With this test setup we therefore measure 59.7 mPas or 6.06% less than the standard value (nominal value). Accordingly, we need to increase all the viscosity data measured by this method by a value of 59.7 mPas to obtain the apparently correct viscosity value. But caution is advised. So far, we have only considered the calibration standard with the lowest viscosity.

In Table 10.2 we see that the standard deviation ranges from 2.25 to 4.65% and is therefore acceptably small. However, the difference between the measured and standard viscosity should be noted. The viscosities we measure with this instrument are too low for low viscosities; conversely, above a viscosity of 2000 mPas we measure viscosities that are too high (Fig. 10.7).

It is therefore entirely possible, as described in the example above, that the sign of trueness can depend on the viscosity standard. In this case we should measure even more viscosities in the range from 4000 to 10, 000 mPas.

If we now look at the differences that can arise with different measuring systems, as shown in Fig. 10.8, we find that the reproducibility (precision) of < 2% obtained for all three measuring systems (CP, PP, cylinder) for two different viscosity standards is very good.

Table 10.4. Comparison of four calibration standards for the measured mean viscosity

Nominal mPas	Actual mPas	Difference mPas	Standard dev. %
985	925	60	2.72
2,000	1,925	75	4.65
5,000	5,080	−80	2.25
12,500	13,100	−600	4.08

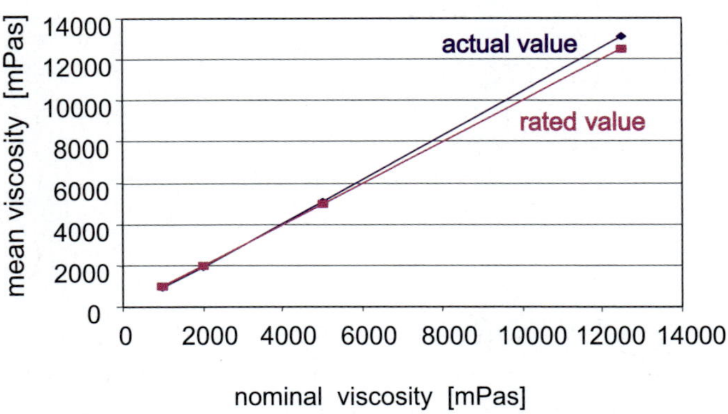

Fig. 10.7. Deviation of the mean viscosities from the standard viscosity

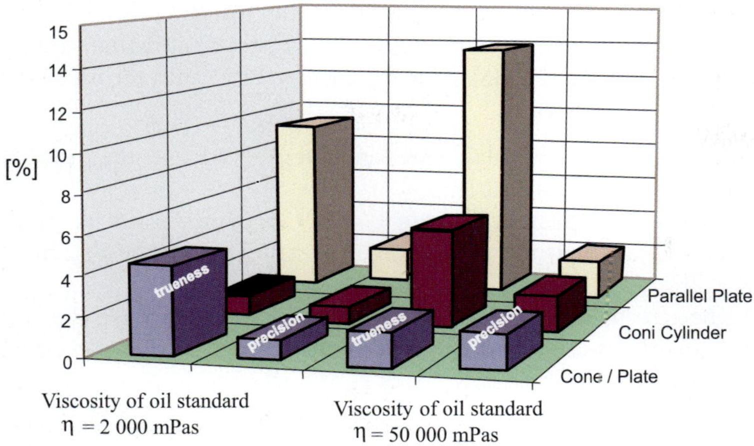

Fig. 10.8. Comparison of measuring systems for the standard viscosity

- The worst results for trueness (i.e. deviation from the nominal value) were obtained with the parallel plate system. This is not surprising as this measuring system is strongly influenced by the boundary conditions.
- The best results for low viscosities were obtained with the cylinder system. This would be expected as well because the geometry of this type of system was developed specifically for this type of sample.
- The cone-plate system can be seen as a fairly universal measuring system for all viscosity ranges, approaching its natural upper limit with increasingly solid samples.

From all the factors influencing calibration/validation measurement, starting with the temperature and the instruments with their respective measuring systems and comparing the respective nominal and actual values, we learn a great deal about our measuring instrument and the problems associated with a viscosity measurement.

How often we need to do such validations or revalidations with just $n = 5$ measurements is a matter for each laboratory to decide for itself. It will also depend on the capabilities and expense. How much is a laboratory head willing to invest in the quality of his results? I sincerely hope it is quite a lot.

If these calibration/validation measurements reveal serious deviations from the previously tolerated result, the instrument manufacturer, i.e. its service, must be consulted.

10.2.1 Calibration Test for Oscillatory Measurements

So far, we have looked at the calibration or validation of viscosity measurements, but what about oscillatory measurements? Naturally, viscosity-related data are inadequate in this case because we are no longer measuring continuously in one rotational direction but sinusoidally around a fixed point. Nonetheless, we can do a calibration test for this type of measurement. Many instrument manufacturers offer a polydimethyl siloxane (PDMS) standard for this. PDMS has the following characteristic properties (Fig. 10.9):

- At low frequencies the loss modulus predominates, $G'' > G' \approx$ the material gets "cold feet" under its intrinsic load.
- At a certain frequency at a temperature of $T = 30\,^{\circ}\mathrm{C}$ both moduli are identical: $G' = G''$. This means that the material is in a state that can be deformed plastically.
- At even higher frequencies the storage modulus G' is greater than the loss modulus G''. In other words this material hardens against increasing external forces and eventually even breaks.

Since crossover (which is what the intersection Gcross is called) only extremely rarely coincides with a measured value, the reported crossover is based on a mathematical iteration. Therefore it is very important always to use the same number and frequencies for this test. The calibration test is considered passed if the crossover measured is in the frequency range of ±5% around the nominal value and the modulus is within ±8% of the nominal value shown in Table 10.5.

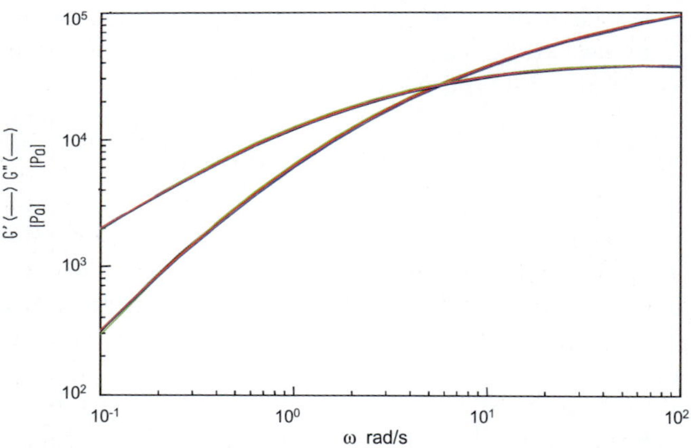

Fig. 10.9. Calibration test with PDMS at $T = 30\,^{\circ}\mathrm{C}$ for oscillatory measurements

Table 10.5. Crossover values for the calibration test

Angular frequency $\omega = 5.508$ rad/s ±5%	Crossover $G' = G'' = 2.7484 \times 10^4 \pm 8\%$
5.2326 bis 5.7834	2.53×10^4 bis 2.97×10^4

10.2.2 Temperature

As we have already seen, temperature plays a crucial role in every rheological measurement. This is why it is just as important to think about checking the temperature and its accuracy.

Users do not need to worry about the accuracy of the temperature because instrument manufacturers have already done that for them. The technology and time needed to control the temperature with an accuracy of $\Delta T = \pm 0.1\,°C$ make control technically feasible but prohibitively expensive for the customer. Therefore instrument manufacturers decided to compromise with an accuracy of: $\Delta T = \pm 1.0\,°C$. This makes it possible to perform most measurements without difficulty in a reasonable (meaning affordable) time.

Instrument manufacturers have also thought about the optimal site for measuring the sample temperature and in most cases have been able to measure it as close as possible to the sample.

How the temperature is attained is another matter. The customer has a number of systems to choose from, starting with a convection oven to so-called heatguns and even Peltier elements or a water bath or air-conditioned room.

If all measurements can be performed at one temperature (e.g. $T = 25\,°C$), the measuring instrument can be set up without temperature control in an appropriately air-conditioned room. If the user wants to measure at different temperatures or even with temperature programs, he already has a difficult choice.

A water bath is no longer state-of-the-art and slow as well. The Peltier element is a very fast and effective system. Discovered in 1834 by the French physicist Jean Charles Athenase Peltier, the Peltier effect can be described as follows: If a current passes through a soldered junction of two metals the junction is heated or cooled depending on the direction the current flows. The Peltier effect is the reverse of the thermoeffect. Today Peltier elements are offered mainly for use as cooling elements (e.g. in camping refrigerators). These elements can, however, also be used to obtain a current from a temperature difference. The disadvantage of this type of heating system is that the sample is heated on one side. This leads to a large temperature gradient especially in thick samples that becomes larger the greater the difference from room temperature.

A smaller temperature gradient can be achieved in the sample with another heating system. Our next task will be to test this system. From the measurement of PDMS at $30\,°C$ we already have indirect proof that the temperature was correct if crossover was found within the tolerance limits.

Elements used for thermal analysis include elements with a defined melting peak such as indium with $T = 156.61\,°C$. This could be used in rheology but the amount of sample needed is very large and also very expensive. The only other alternative is to have the instrument manufactures calibrate temperatures at least once a year.

11 Tips and Tricks

The purpose of this chapter is to give current and/or future viscometer or rheometer users some important advice on how to eliminate in advance many hidden sources of error. The main goal is to make readers more sensitive to avoiding errors in order to be able to perform quick and reliable measurements.

11.1 Materials for Geometric Systems

What materials should measuring systems be made of? Primarily four materials are used: stainless steel, titanium, acryl and aluminum.

- Stainless steel is used mostly for corrosive samples. The material is additionally characterized by high inertia and reduces thermal gradients at high temperatures.
- Titanium is suitable for special demands such as a high pH combined with low inertia.
- Acryl is used mainly for weak gel structures or dilute solutions (not organic solvents) up to 40 °C.
- Aluminum is used primarily for replaceable, meaning disposable, measuring systems and suitable for all samples with the exception of those with a high pH.

For the correct material for the measuring system, we need to think about the correct geometry for the product we want to measure. This must be adapted to the viscosity of the sample. When should we use a cone, a plate or a cylinder system?

11.2 Cone-plate

This measuring system is suitable for all bulk liquids and dispersions (suspensions and emulsions) [96] with a particle size no larger than 1/5 of the virtual gap.

There are a variety of diameters to choose from:

- For low viscosities like oils a large cone diameter (e.g. 5 cm)
- For medium viscosities like honey a cone with a diameter of 4 cm
- For high viscosities like flan (cream caramel) a 2.5-cm cone diameter

The following rules should also be remembered.

- The smaller the cone diameter, the larger the shear stress
- The smaller the cone angle, the smaller the shear rate to be measured
- A cone angle of 1 degree means a virtual tip of ca. 30 µm
- A cone angle of 2 degrees means a virtual tip of ca. 60 µm
- A cone angle of 4 degrees means a virtual tip of ca. 120 µm

11.3 Parallel Plate

This measuring system can be used for all filled samples with particle sizes too large for measurement in the cone-plate system:

- The gap should be between 1 and 2 mm
- If the gap is smaller than 1 mm, the measured viscosity will be too low
- However, if the gap is too large (> 2 mm) this will lead to a thermal gradient in the sample

Especially in the case of systems heated on one side like the Peltier element the temperature difference ($T_{\text{Peltier}} = 50\,°\text{C}$) in a sample 2 mm thick can be $\Delta T = 10\,°\text{C}$. This means (Fig. 11.1) that the temperature at the heated bottom plate is $T = 50\,°\text{C}$ but at the top unheated sample the temperature is not more than $T = 40\,°\text{C}$ even after a waiting period of over 20 min.

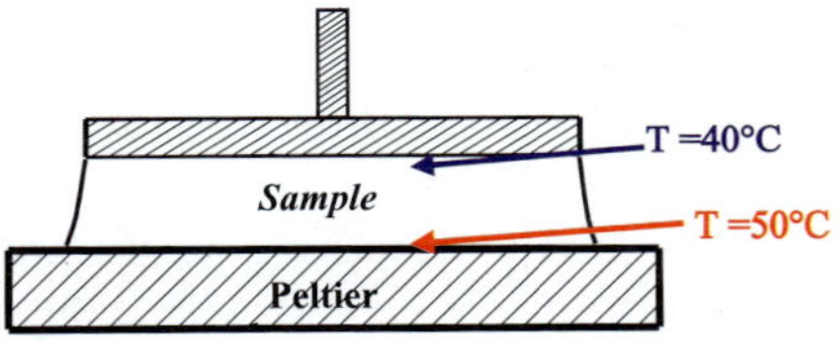

Fig. 11.1. The Peltier element generates a temperature gradient

11.4 Cylinder Systems

These are suitable primarily for liquids and suspensions. Double gap cylinder systems are ideally suited for measurements with minute shear rates. For these systems it must be ensured that no air bubbles become trapped in the measuring gap, as this would falsify the measuring result.

11.5 Cleaning Measuring Systems

The cone-plate and parallel plate systems are the easiest to clean. This can usually be done with a soft cloth and a liquid like water or a solution of isopropanol and water. The measuring systems should never be damaged mechanically. This may result in a plate that is no longer plane-parallel and rotates slightly off-center. If the edges are damaged carelessly, the whole system must be replaced by a new, intact system.

The design of the cylinder system makes it difficult to clean because the corners of the outer cylinder are hard to reach. The only solution is to be patient and to remove all residues by rinsing several times. Also available are expensive models in which the lower base can be screwed off.

11.6 Measurement Artifacts

- Viscoelastic fluids frequently leave the measuring gap. The viscosity values are noise-infested or too low. Solution: use of a smaller gap or angle to achieve the same shear rate at a lower angular velocity.
- In the cone-plate and parallel plate systems the viscosity may decrease dramatically at low shear rates (Fig. 11.2) because the sample is slung from the gap by high centripetal forces. Slinging of the sample material from the measuring gap leads to a decrease in the viscosity data because less material is available for measurement. If we still need to record viscosity at high shear rates then we will just have to choose another instrument like the high-pressure capillary viscometer.
- Wall slip arises when there is a tendency for a lower-viscosity phase to separate from the sample material. This problem can be solved to a certain extent by using other plate materials or roughened plates but also by using another measurement method.
- Steady-state flow is associated with long measuring times for low shear rates. When combined with drying out of the sample, this can in turn result in the need for several measurements, each requiring new sample material, to obtain a meaningful flow curve.

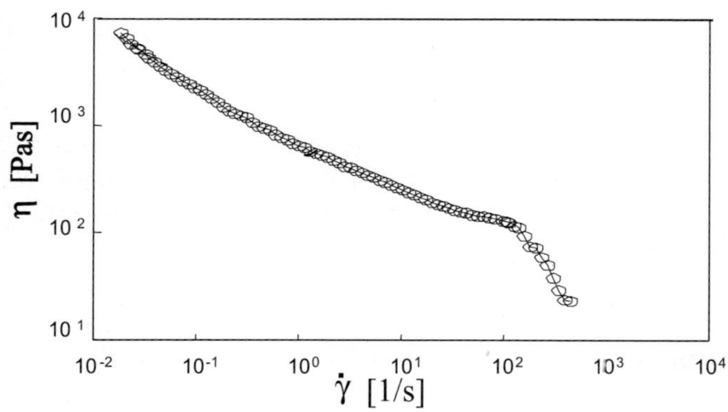

Fig. 11.2. Slinging of the sample from the measuring gap

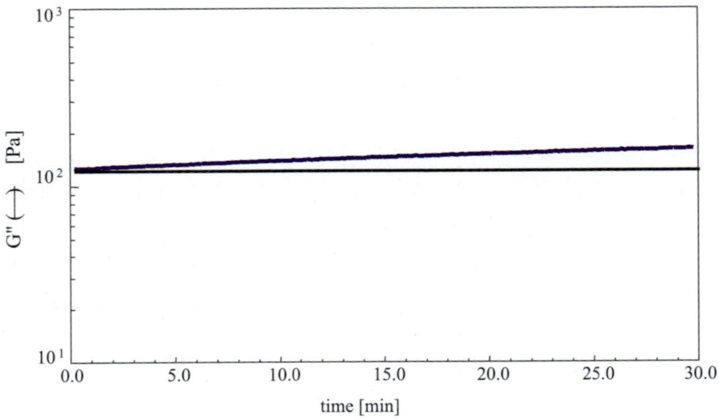

Fig. 11.3. Increase in the storage modulus G' due to drying

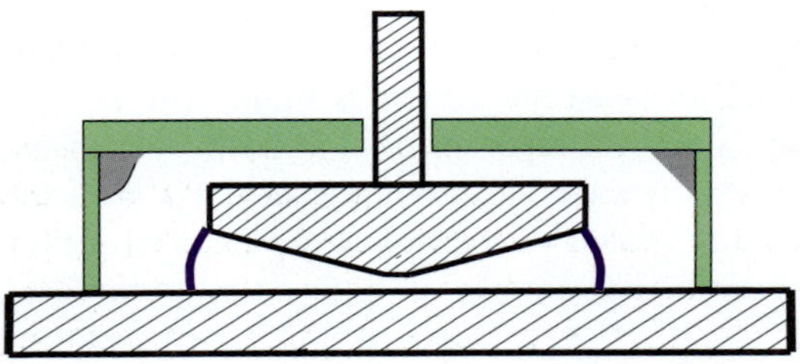

Fig. 11.4. Solvent trap over a cone-plate system

- The Cox Merz rule [85] known from polymer solutions states that the viscosity data from dynamic frequency measurements are identical to those from shear rate measurements in the linear measuring range. This rule does not apply to dispersions (exceptions possible).
- Drying out at the edge of the sample results in a viscosity increase. This can be recognized by an increase of G' in a time sweep (Fig. 11.3).

One solution can be to change the geometry. Although a large plate will increase the free lateral area for the sample, a reduction in size will take effect relative to the total sample volume.

A solvent trap like that shown in Fig. 11.4 is the better solution. A cap-like construction is placed over the sample without touching the measuring system. Found in this cap are small sponges filled with a liquid. This creates a moist atmosphere for the sample.

11.7 Filling of Cone-plate and Parallel Plate Measuring Systems

Sample preparation is crucial to a measurement [97]. First the measuring systems need to be preheated to the desired temperature. This can take several minutes depending on the temperature. Then close the gap between the parts of the previously cleaned measuring system. In the parallel plate system, these are the plates. This procedure is called zero adjustment. The zero adjustment must be performed every time the measuring system or temperature (isothermal measurements) is changed. This can be done manually by slightly rotating the upper plate and slowly reducing the distance between the plates. The zero position is reached when the slightly rotating upper plate comes to a standstill. For many instruments, a suitably pre-programmed zero adjustment eliminates the need for this step.

Afterwards separate the plates (this should be done very quickly) and place the sample to be measured in the center of the lower plate. When filling any measuring system be sure to avoid putting a large strain on the sample due to filling.

- Therefore do not use small syringes that produce high strain and can result in loss of structure.
- Cut open tubes and remove the sample carefully for instance with a spatula. The small opening of the tube acts like a capillary and already stresses the sample in advance.

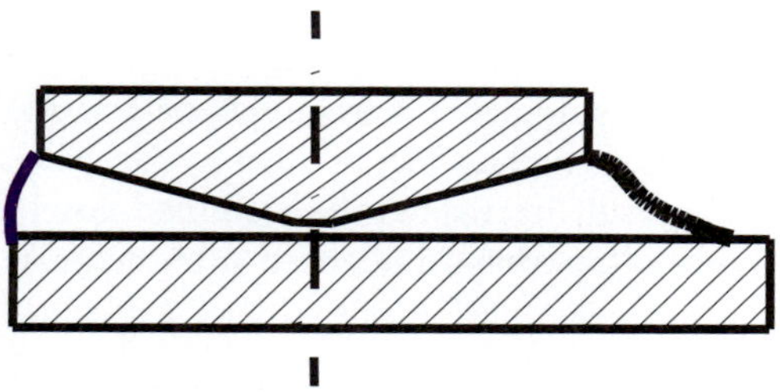

Fig. 11.5. Correctly filled cone-plate system (*left*)/overfilled (*right*)

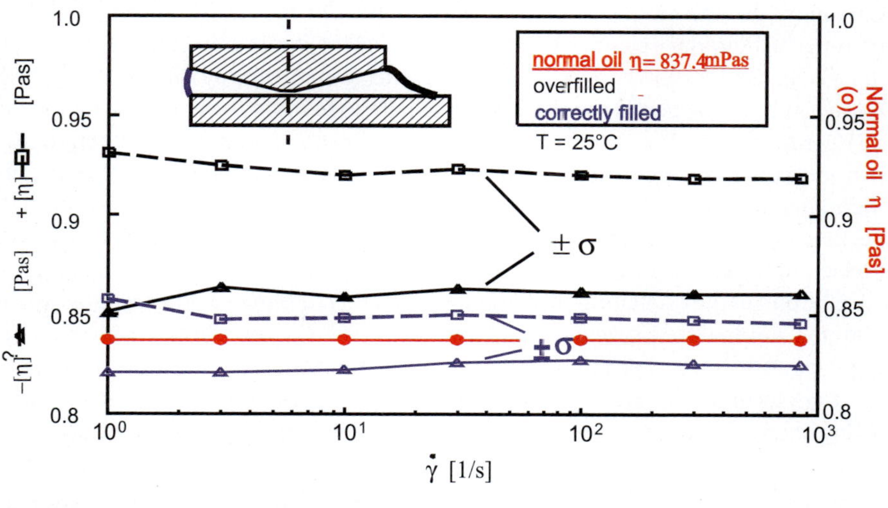

Fig. 11.6. Measurements with gaps that are correctly filled and overfilled

- Special measuring spoons that also provide the correct sample amount are particularly suitable.
- Close the gap slowly and stop approximately 50 µm before reaching the measuring gap height and clean the edge (Fig. 11.5).
- Then slowly close the gap the remaining 50 µm.
- Never change directions while closing the gap.

It can be clearly seen in Fig. 11.6 that the mean increases with overfilling and that the scatter, meaning the standard deviation, is also significantly larger than with correct filling. These two effects are easily explainable. Since the amount of sample in the measuring system is not constant when the measuring gap is overfilled, the scatter is greater. The mean inevitably increases because more sample material than actually intended is measured by the sensor. Similar but opposite effects are obtained when too little material is placed in the measuring gap. The mean viscosity decreases and the scatter increases slightly relative to a measurement with a correctly filled gap.

11.8 Interpretation

The first rule in interpreting rheological measurements is to be self-critical and determine whether the result as presented is even plausible. In other words, when interpreting rheological measurements the result first needs to be examined closely. Does it meet our expectations or were the results perhaps influenced by the artifacts described above?

Found in the literature are many models that can be used to describe viscosity functions. They are helpful especially when the samples to be described are very similar or identical. Usually, however, one sample can be described better with the Casson model and another with the Herschel–Bulkley model. This is why, besides naming the boundary conditions of the measurement, it is also so important to state the model used for interpretation. If we want to compare many different samples, however, it may be useful to refrain from using models and to concentrate entirely on the raw data. For example differences can be clearly recognized by superposing two measurements because the human eye can discern very fine differences. For the interpretation of flow limits we feel that no models should be used and that best results are obtained by plotting the viscosity vs. the shear stress as explained in Sect. 7.1 (Fig. 7.10).

The plot of the oscillatory measurements in Fig. 11.7 shows a difference between both the storage and loss modulus of approximately 11% for the two samples. The question now is whether this difference is significant.

To determine this we should consider another parameter, the loss factor tan δ (Fig. 11.8). There we find clearly smaller differences of approximately 1.5–2% because forming the quotient of the loss factor eliminates the effects of geometry produced by normal forces that are not always avoidable.

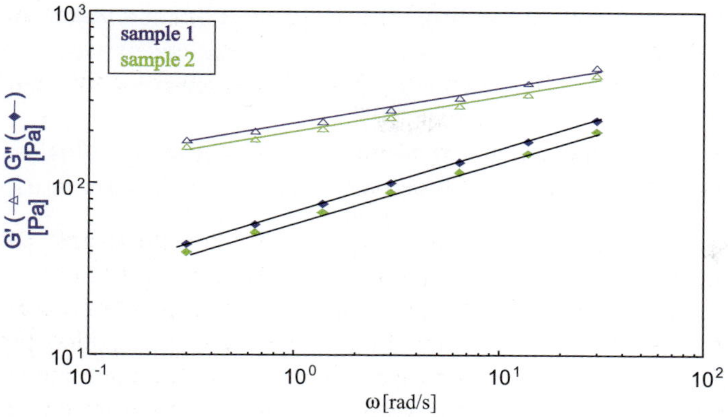

Fig. 11.7. Are the differences between the two samples significant?

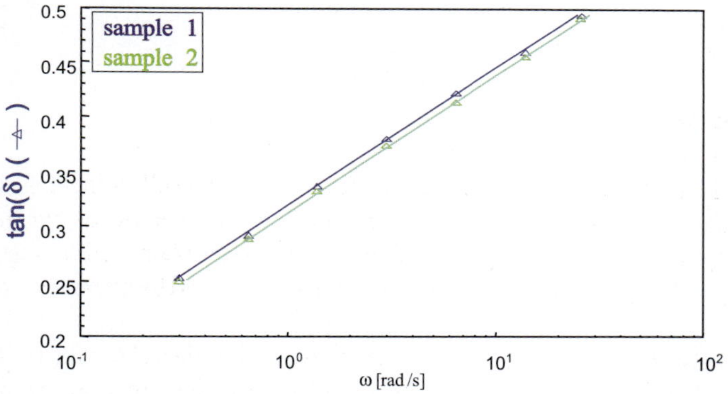

Fig. 11.8. Loss factor $\tan \delta$ for two samples (with no influences from geometry)

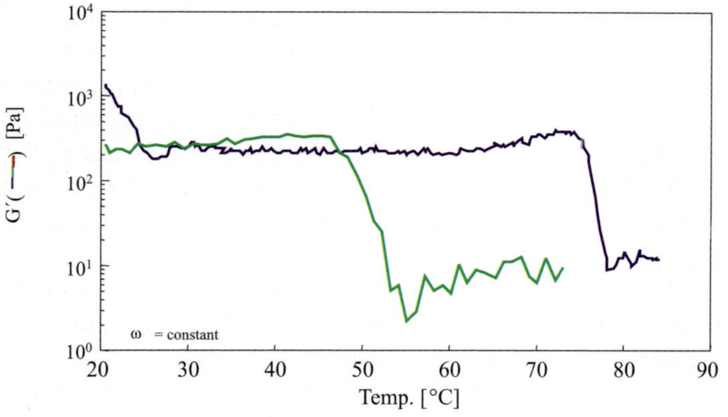

Fig. 11.9. Difference in the temperature dependence of two samples

The results indicate that both samples should be classified as identical rheologically. To confirm this either a frequency test should be performed at two more temperatures or a temperature test at one frequency in a larger temperature range (Fig. 11.9).

Here we can clearly see that although both moduli are the same at 25 °C, distinct differences arise at a higher temperature. However, since both curves are nearly identical, in all probability the samples are identical.

12 Definition of Cosmetics

Before we look beyond the area of cosmetic emulsions, it might first be appropriate to point out the differences between a cosmetic product and a drug under German law.

12.1 Cosmetics vs. Drugs

Cosmetics are (according to Brockhaus, as defined by the German Food and Commodities Act of 08/15/1974) [98] substances or preparations made of substances intended mainly for external use on the human body or in the oral cavity for cleaning and personal hygiene to alter the appearance or body odor or to convey scent.

The definition is therefore based first on the intended use of the product, but this is not always the case. If a product is a cosmetic according to its intended use, we still need to ask whether it could also be used to influence disease. On this point a distinction is made between cure, alleviation and prevention. If a product is used mainly for the cure and alleviation of disease, it is a drug even if, in addition, a secondary cosmetic purpose is pursued. If the product serves mainly to prevent disease, it is a cosmetic even if the cosmetic purpose is secondary.

But there are always exceptions: this distinction does not apply if the product is intended to alter the shape of the body, i.e. the size of the body. The reason for this strict classification as a drug is the "main" intervention in body functions by products with a body-shaping effect.

When determining the purpose served by a specific product, consideration is given to the description, labeling, advertising and presentation of the product but not to the actual suitability of the product – as a rule this will remain unknown to the non-expert consumer.

12.2 Production of Cosmetic Products

Anyone who produces cosmetic products or has them produced in his name is a manufacturer and must fulfill the appropriate obligations. When formulating the exact composition of a cosmetic product the Cosmetic Directive must be consulted. It regulates both harmless additives and maximum quantities.

Under section 5d paragraph 1 of the Cosmetic Directive the manufacturer [99] must meet information requirements:

- Before the cosmetic product is marketed for the first time, the appropriate authority must be notified of the places of manufacture.
- At the same time the Federal Institute for Consumer Protection and Veterinary Medicine must be notified of the trade name, product description and composition of the cosmetic product. This Institute may only use this data for treatment of possible health damage.
- At the same time the manufacturer must keep documents available at the place of manufacture covering the cosmetic product. These documents must in particular record data on the composition, the safety assessment and the name and address of the person responsible for the safety assessment.
- A responsible person must be determined for the cosmetic product's safety assessment as regards human health; this person must assess the cosmetic products in line with the principles of good laboratory practice. The precondition for this is that the responsible person has a degree in pharmacy, medicine or a similar professional field.

12.3 Naming, Trademark Law, Patents Law

When naming any product it must always be ensured that other manufacturers' existing trademarks and naming rights are not infringed. It must also be ensured that the product or parts of the product do not infringe other patents. If this is ignored, there is the risk of facing warnings with costs and high legal expenses and also bans preventing the marketing of products which have already been manufactured.

12.4 Marketing of Cosmetic Products

The Cosmetic Directive also contains regulations covering the labeling of and special information on products – such as marks to identify the manufactured lot, minimum best-by date, ingredients, etc. These data must be on the container and on the packaging. If this is not possible due to size, reference must be made to a package insert. Currently, a change in the demands for minimum shelf-life data is mainly under discussion (e.g. 33 months after opening the pack).

Special care must be taken to ensure that the products in no way damage health when used as intended or foreseeable. Foreseeable use can be judged by the presentation, labeling and all the other data accompanying the certificate. The manufacturer can only be advised to indicate warnings prominently and clearly on the product and its packaging because irrational behavior by the consumer can also be "foreseeable", e.g. ignoring or overlooking enclosed instructions for use.

No special requirements apply to cosmetic products sold in normal trade channels (supermarkets, drugstores). This situation changes when a doctor is involved in product marketing. No matter where cosmetic products are sold and independent of the doctor's contribution to the manufacture and marketing of such products, the independence of the medical decision must always be ensured. Section 34 paragraph 1 MBO (Model Professional Regulation for German Doctors) [100] therefore bans any contract wording which makes the doctor's remuneration dependent on what products he recommends.

The doctor is not permitted to hold promotional lectures for cosmetic products or to compile specific expert reports for advertising purposes (section 34 paragraph 3 MBO).

Product sales in the doctor's office are substantially restricted by section 3 paragraph 2 MBO. This prohibits the doctor from dispensing or actively having dispensed goods and other objects while carrying out his medical activities. The situation is only different if such distribution is a necessary part of the medical therapy. This cannot apply to cosmetic products, however, because they normally cannot be part of medical treatment. For this reason the marketing of cosmetic products in the doctor's office is as a rule inadmissible. This applies both to marketing by the doctor and his assistants and to marketing by third parties who have been granted access to the office by the doctor.

It is possible to market cosmetic products in a cosmetics institute in which the doctor can certainly have an economic interest.

12.5 Advertising Cosmetic Products

The restrictions under the Law Prohibiting Unfair Competition (UWG) [101], the Food and Commodities Law (LBMG) [102] and possibly the Drug Advertising Law (HWG) [103] must be observed when advertising cosmetics.

Anyone infringing the UWG can be taken to court in an action for an injunction and for damages by competitors and consumer organizations. When an advertising activity infringes the UWG can, however, often not be determined exactly. The general clause, section 1 UWG, prohibits every competitive activity which offends common decency. The campaigns this covers have to be determined by the courts in individual cases.

Section 3 UWG prohibits misleading data. In the field of cosmetic dermatology this regulation is chiefly significant as regards price data and similar data concerning economic details. Where statements on the effects of cosmetic products, processes and treatments are made, they must be tested according to the stricter HWG and substantiated by the manufacturer with appropriate proof.

In addition, the UWG also contains regulations on comparative advertising, on wholesale sales to end consumers and progressive customer solicitation.

Finally, it must be said that the doctor or company can also be prosecuted under the UWG if – in any anticompetitive way – they disregard other standards such as the HWG, professional regulations, etc.

The HWG covers numerous products and services in so far as they (also) serve to cure or alleviate illnesses.

Under section 1 HWG the law also applies to drug advertising and to the promotion of other products, processes, treatments and objects. "Other products" also include cosmetic products.

The basic demands made of advertising can be found in section 3 HWG. This regulation mainly applies only to advertising within expert circles because advertising aimed at laymen is further restricted by section 11 HWG.

Section 3 HWG:

Misleading advertising is inadmissible. Misleading advertising is in particular

1. When drugs, processes, treatments, objects or other products are claimed to have a therapeutic effectiveness or effects which they do not have
2. If the impression is erroneously given that
   a) Success can definitely be expected
   b) No harmful effects will occur, if used as stipulated or for a longer period
   c) The advertising is not used for competitive purposes
3. If untrue or deceptive data are provided
   a) About the composition or nature of drugs, objects or other products or about the nature of the processes or treatments
   b) About the person, previous experience, competence or successes of the manufacturer, inventor or the persons who act or acted for them

Inadmissible under section 6 HWG is advertising when unprofessional and hardly checkable reports, certificates or publications are used for promotion purposes. In addition, quotations, tables or other material taken from specialist literature must be reproduced accurately.

Even stricter regulations apply to advertising outside expert circles. Under section 11 HWG the following means of advertising are inadmissible – over and above the demands contained in section 3 HWG:

- Expert reports, certificates and publications; letters of thanks, recognition or recommendation
- Statements that the drug or process is recommended or checked or used by professionals
- Details of patients' medical history
- Pictures of people in white coats or working in medical professions or as pharmacists
- Pictures of diseases or physical injuries, in particular before/after pictures and pictures of the effects on a human body of a drug or process
- Descriptions in foreign languages or of a technical nature which are incomprehensible to the average reader
- Causing fear
- Offering or accepting addresses
- Disguising the advertising purpose
- Encouraging self-diagnosis and self-treatment

- Addressing the target group of children under 14
- Contests, raffles
- In the case of drugs, every free handout of samples; in the case of other products or objects only handouts not asked for by the customer

In normal business dealings with the customer giveaways – i.e. samples, test treatments, etc. – only low-value trifles can under sections 6 and 7 HWG be handed out. The HWG is somewhat more generous as regards advertising within the framework of exclusively work-related scientific events.

In addition to the advertising restrictions which apply to cosmetic products under the UWG and the HWG there are also special demands contained in section 27 LMBG, demands which admittedly overlap to some extent with these basic demands. Particular and independent significance is gained by the LMBG regulations when cosmetic products serve only to prevent disease or only aesthetic purposes; in such cases the HWG advertising restrictions are not applicable.

Under section 27 LBMG it is not permitted to market cosmetic products using a misleading description or presentation or to advertise using misleading information. According to this, misleading information is given when effects are attributed to cosmetic products but their scientific certainty is inadequately proven. In par-

Fig. 12.1. Justicia

ticular, it is prohibited to create the false impression that success can be expected with absolute certainty. Also prohibited are descriptions and statements which can be deceptive as regards the manufacturer, inventor, origin, quantity, shelf-life and other circumstances which influence the evaluation of cosmetic products.

12.6 Comments

Cosmetic products must be clearly distinguished from drugs. The regulations which apply to each product group differ too greatly, in particular in the areas of production, registration and monitoring.

Despite the basically different legal standards there are numerous points of contact in the regulatory system. In many cases advertising law applies the same standards to cosmetic products and drugs. As the consumer sometimes finds it difficult to distinguish the two, equal treatment is appropriate. But here too the following applies: only a definite distinction between the two will allow nuances in the application of law and jurisdiction to be determined.

13 Excursion in the World of Food Rheology

The way substances flow or resist a body passing through them has been the subject of interest and observation [104] from the earliest times. It is fairly safe to say that foods belong to substances with textures, as we would say today, that were evaluated from a rheological viewpoint. Therefore we want to widen our horizons from the food rheology point of view. But first we will take a short peep into the past.

13.1 A Short History of Food Rheology

For instance Lucretius described the flow of substances as follows:

"... For water moves and is made to flow by the slightest force because it is made of little, rolling particles. In contrast, honey is more stable, its flow more sluggish and its movements slower, for it has an internal cohesion. The likely reason is that it is formed of particles that are not so smooth nor so fine and round".

Some names of foods are derived from their rheological consistency. One example is the name Molle (Lat. mollis = soft) [105] used for breadcrumbs in the Tyrol. The German word "Schmer" used for butter and lard indicates the rheological properties that describe spreading (smearing). Everyone involved in preparing food, whether a baker, cook or housewife, knows how slight changes in the method of preparation can produce foods and dishes with special rheological properties. Boiling, baking and roasting not only change the consistency, they also produce the typical flavor of a food. Flour is used in every kitchen to adjust the consistency of foods (e.g. thickening of gravy).

Even today we still use our hand, which Kant called the visible part of the brain, as a rheological measuring instrument. There is evidence that bakers in ancient Egypt judged their dough by rolling a piece of it back and forth between their fingers. A picture from the time of Ramses III (2000 BC) shows the daily routine in a bakery. Two bakers can be seen kneading dough with their feet and holding rods in their hands, which they are obviously using to check the firmness of the dough (Fig. 13.1).

Instructions in recipes in old cookbooks often contain rheological terms. Typical examples for dumplings are: "The dough has the correct denseness when it separates from the bowl and spoon." "Gently knead the dough without stirring to ensure the liver dumpling dough does not become too firm."

Fig. 13.1. Bakery in ancient Egypt

The following example [106] of the exact description of the flow properties of a sugar solution on heating by the confectioner shows that differences in the rheological properties of raw materials have long been known.

There are eight degrees of sugar boiling:

1. To small thread
2. To large thread
3. To small pearl
4. To large pearl
5. To the blow
6. To short feather or soft ball
7. To long feather or hard ball
8. To crack or caramel

The test for the first level is described as follows: "Boil the sugar with stirring. To sample remove the spoon from the pot, spread a little of the sugar on your index finger and draw it into a thread with your thumb. If the sugar forms a thread that immediately breaks and leaves a drop on your thumb it has been boiled to the thread stage. If the thread is hardly noticeable, the sample is a small thread. However, if the thread is drawn out somewhat farther without breaking and becoming brittle, the sugar has been boiled to the large thread stage." To identify the eighth stage we need a pot of fresh water containing a smooth, round rod. Using this rod or your finger, which was previously dipped in cold water, take some sugar and put it immediately in cold water. If the sugar separates from the rod or your finger and cracks, it has been boiled to the crack stage, as the name implies. If this level has been reached, continue to test it as follows: "After the sugar has cooled in water put it between your teeth. If it sticks like tar to your teeth it has still not been boiled to caramel and needs to be boiled again several times. Afterwards, as soon as you put the rod with the sugar in cold water the sugar will crack and no longer stick to your teeth."

13.1.1 The Origins of Food Rheology

In the middle of the last century food science originated as a field of applied chemistry. Later hygienic aspects were added. In the process [107], knowledge of the chemistry and microbiology of food grew thanks to systematic studies on extensive fields of science. The field of food rheology, however, developed only slowly and almost in secret. Nevertheless, a very few scientists and practitioners devoted themselves for very different reasons to rheological topics:

- It was already observed fairly early that not all raw materials are equally well suited for manufacturing foods and that a variety of rheological properties could be the cause. Bread dough from flours of different origin resulted in different consistencies that could be determined from the feel of the dough.
- At the end of the 19th century scientists like Bingham began to study the rheological properties of colloids. They quickly recognized that important properties of foods are determined by their colloidal nature and that colloid-chemical processes also play an important role in food processing.
- However, sensory properties like the chewability of foods can also vary and be influenced by the production process.
- The “mouthfeel” is a criterion used in ice cream tasting to evaluate the texture and melting behavior. A variation in the production process can once again cause changes in these properties.

The following examples should demonstrate that rheology is an important tool not only in product development for cosmetic emulsions or polymers but also, consciously or unconsciously, in daily life. On the following pages we will be considering several examples relating to food rheology. Once again we will start with everyday things that are a part of daily life and look at the properties of such products as honey, butter, cheese or margarine. But we will also take a closer look at psychorheological aspects of these products.

13.2 Honey

The first rheological studies on honey were performed to detect adulterated products containing sugar or starch syrup. Using the Ostwald viscometer it was determined that starch syrup additives increased the viscosity of honey considerably due to the dextrin content. Every German housewife [108] knows that gelled honey from the German Heide can be made pourable again by stirring.

This behavior of honey from the German Heide can be described rheologically as an isothermal reversible gel-sol-gel transformation after shear forces were applied and the gel was subsequently left undisturbed. This thixotropic behavior (Fig. 13.2) could also be detected for other types of honey. Normally honey behavior is Newtonian-like.

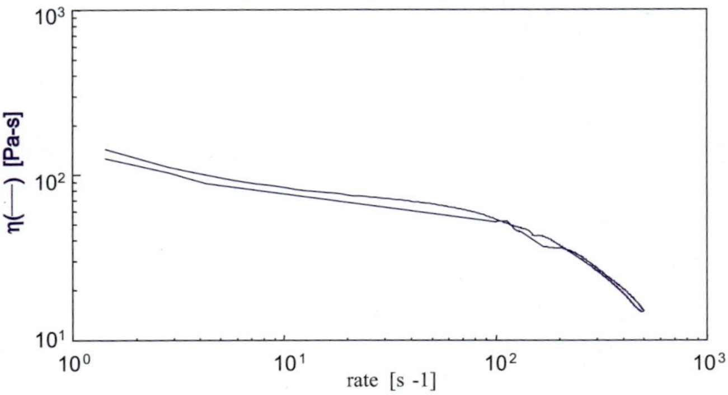

Fig. 13.2. Thixotropic behavior of German Heide honey

13.3 Sandwich Spreads

Butter was first subjected to rheological measurements to find out if there is a relationship between the melting point and hardness of butter. As usual, it was found that there are always exceptions to the rule. Depending on how butter is cooled, it can be hard despite a high melting point [109] or soft despite a low melting point.

A much more interesting problem from a rheological-sensory standpoint first described in 1949 is the spreadability of butter. In Fig. 13.3 the different rheological properties of two different spreads are clearly recognizable. The storage modulus G' is nearly the same for both products in the measured frequency range and at $T = 25\,°\text{C}$. This is not surprising, as both products are supposed to have a similar texture at room temperature. However, the curves for the loss modulus G'' clearly reveal two different products. It can be seen that the values for Sample B are shifted

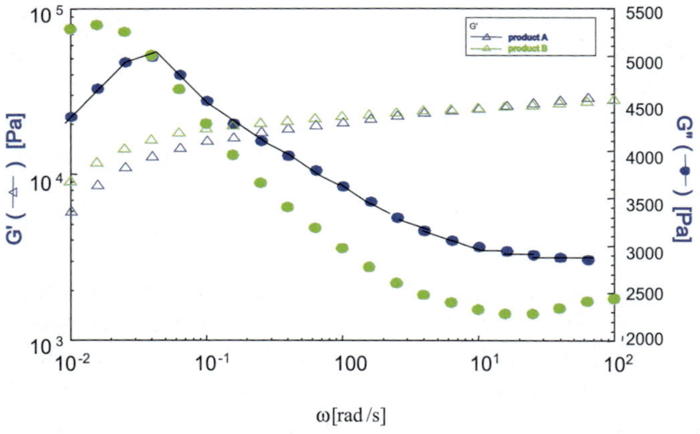

Fig. 13.3. Oscillatory measurements of sandwich spreads

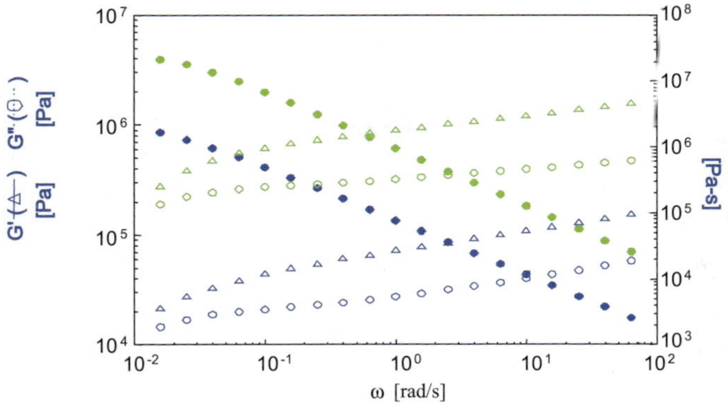

Fig. 13.4. Frequency test for two cheese varieties

to lower frequencies. One explanation might be a different raw material vendor or a deliberation change of a raw material.

13.4 Cheese

An important step in the production of most kinds of cheese is the conversion of liquid milk to a gel-like curd by exposure to LAB [110] (lactic acid bacteria). The experienced cheesemaker knows how firm the curd must be for production of a specific cheese variety.

To evaluate the firmness of curd an instrument is still used today that was developed by Allemann and Schmidt in 1722. Three copper wire rings 3, 5 and 7 cm in diameter respectively were soldered concentrically onto two intersecting wires. The wire grid was pulled through the coagulated milk at constant speed by a string and the resistance measured by a spring scale. This method was used to perform the first rheological measurements of the effect of different amounts of LAB on the firmness of curd and the dependence of the firmness of curd on the pH and temperature [13]. The rheological properties of two finished cheese varieties are presented in Fig. 13.4.

13.5 Ketchup

Ketchup [111] is an English word from the Malaysian word kechap. It means literally "spicy fish sauce". British colonialists took back (it is uncertain whether it was from Indonesia, Malaysia or Thailand) a variant of this (spicy) sauce made of preserved fish to Great Britain where it was modified in the 18th century. Today the main ingredients are tomato paste, onions, vinegar, spices and sugar. The Stiftung Warentest (German Consumer Reports) found up to 29% sugar in tomato ketchup in 1997.

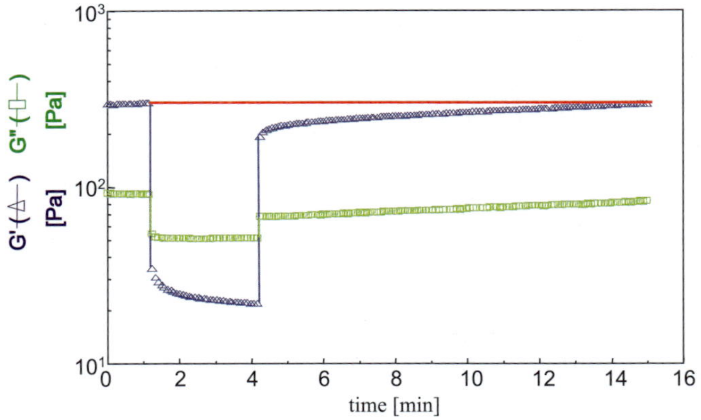

Fig. 13.5. Structural breakdown and buildup of tomato ketchup

We all know that ketchup is red and needs to be shaken before use. To get it out of the bottle it first needs to be shaken vigorously for a few seconds and the contents of the bottle become firm again when left to stand undisturbed. This behavior is said to be thixotropic. The structural breakdown and buildup shown in the rheological measurement in Fig. 13.5. represents this behavior. The first part of the measurement is performed in the linear viscoelastic range. In the second part the strain is increased sharply. This results in a higher value for the loss modulus than for the storage modulus. In the third part we return again to the viscoelastic range, and after a few minutes the moduli have again reached their starting values.

13.6 Yoghurt

Yoghurt [112] has a structure and firmness similar to that of ketchup. Stirring causes this product to flow, but the question remains: is yoghurt thixotropic?

The answer is NO! We did the same structural breakdown/buildup test with yoghurt as with ketchup. It could be clearly recognized (Fig. 13.6) that due to the greater deformation the absolute values of both moduli G' and G'' decreased sharply and the loss modulus G'' became larger than the storage modulus G'. After the sudden return to the smaller deformation the relationship reversed very quickly but the values never reached the baseline condition.

13.7 Marzipan

Marzipan [113] has been known in Europe since the beginning of the 15th century. According to philologists the word "marzipan" comes from the Arabic word "maulhaban" meaning "reigning king". The likeness of a reigning king was imprinted on a small flat cake made of chopped almonds and sugar. This small cake was packed

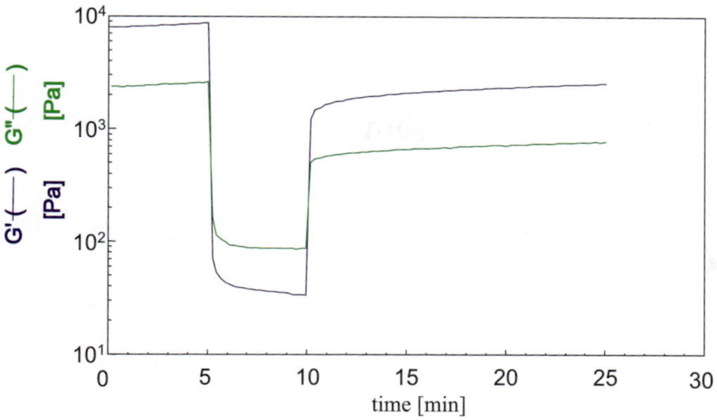

Fig. 13.6. Non-thixotropic behavior of yoghurt

individually in a specially prepared box and sent to Cyprus. From here it reached Venice via the trade routes.

From Venice marzipan spread throughout the world. Legend has it that marzipan was created in 1214 in the San Clemente Closter in Toledo. A famine gave some nuns the idea to gather almonds from the trees on their grounds, grind them and make a paste that resembled bread dough. To make the almonds less bitter sugar was added to the paste.

The basic ingredients of marzipan today are still almonds and sugar, but many ingredients have been added and special processing techniques developed. According to German law products marketed under the name marzipan may have a mixing ratio of 50 parts marzipan mass to 50 parts sugar. In Fig. 13.7 the behavior of a marzipan mass at different temperatures is shown. The $\tan\delta$ curve at 25 °C shows distinctly better processing conditions because it is nearly constant in the measured frequency range and at values of $\tan\delta < 0.35$ still has a significant viscous portion at high frequencies.

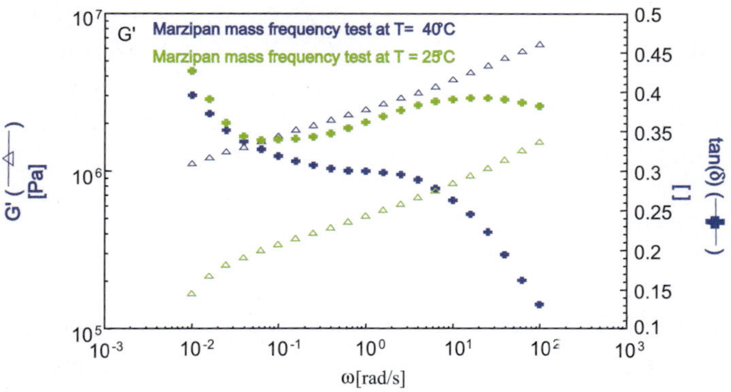

Fig. 13.7. Marzipan mass at two temperatures

13.8 Starch

Wheat starch [114] was used in ancient Rome as a food thickener. The dependence of the viscosity of starch pastes on the temperature was studied by Wolfgang Ostwald. He called the temperature at which the viscosity suddenly sharply increases the gelatinization point (Fig. 13.8).

To determine the breaking force and firmness of starch gels instruments similar to those used to study gelatin and pectin gels are employed. Of the many instruments described for measuring the rheological properties of starch pastes only the viscograph has established itself in actual practice. It is a further development of the amylograph and can be used to determine the viscosity also in the cooling phase of starch pastes.

As Fig. 13.9 shows, however, even a simple viscosity measurement reveals dilatant behavior, which otherwise occurs only very rarely in nature. The viscosity increases with increasing shear rate.

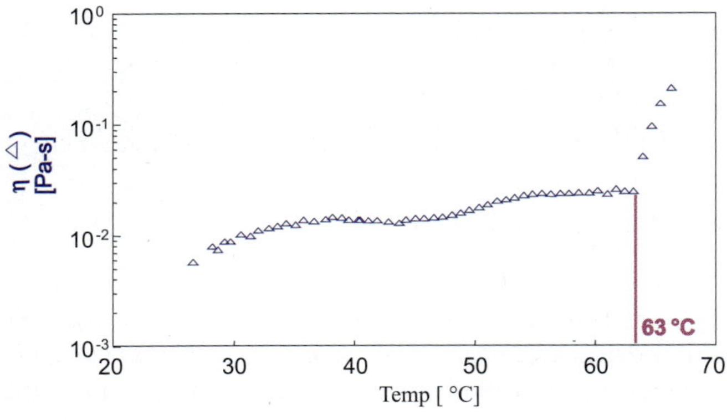

Fig. 13.8. Gelatinization point of a wheat starch

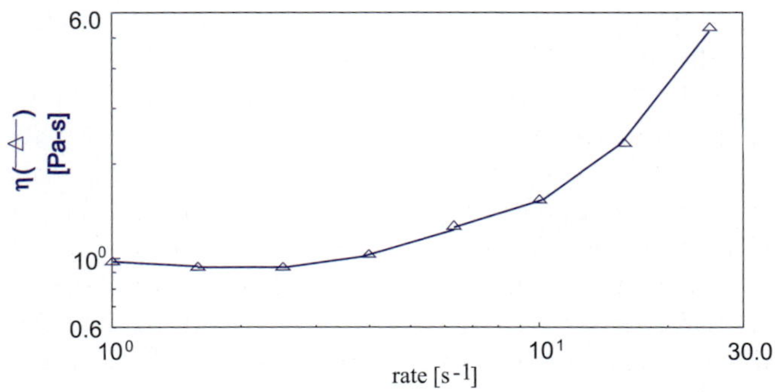

Fig. 13.9. Dilatant behavior of wheat starch

13.9 Foams

Many foam dessert products [117] are found in the food industry, with ice cream, protein/sugar foams like marshmallows, whipped cream and chocolate mousse being some of the best known examples.

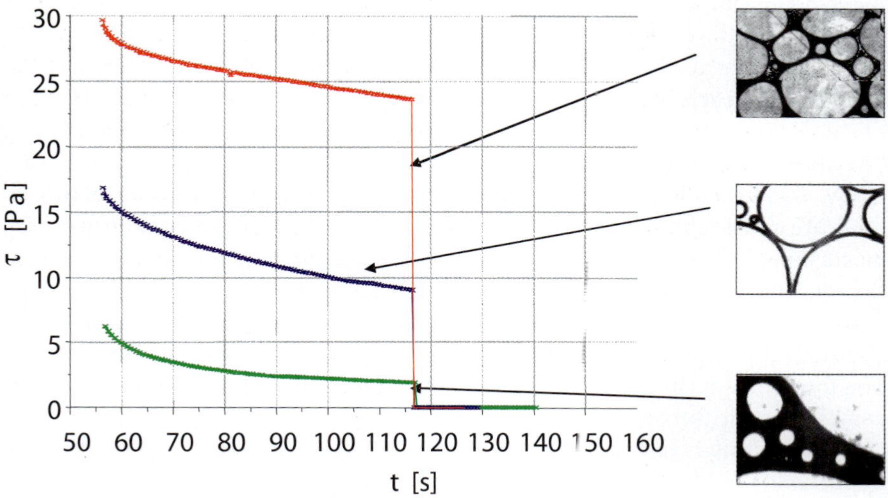

Fig. 13.10. Combined rheo-optic measurements on foams

A foam is a liquid or solid-like substance in which gas is finely dispersed. The gas bubbles are stabilized by surface-active substances. Normally, food foams are produced with continuous rotor/stator whipping machines in which turbulent flows prevail. The most important process parameters are the speed of the rotor, the holding time in the whipping head or throughput, the temperature, the geometry of the rotor/stator discs and the pressure in the whipping head. Currently, one of the focuses of our research is to study more closely the influence of these parameters on the microstructure of the resulting foams. The aim is to use this information to improve further the production process/whipping machine to obtain as fine-bubbled foams as possible with a narrow distribution of bubble sizes. The main advantages resulting from small and narrowly distributed air bubbles are improved texture (the so-called mouthfeel is creamier), improved flow properties and a longer shelf-life of the finished product. These dependencies not only apply to food foams but also to synthetic polymer foams as well as cosmetic and pharmaceutical foams.

Analysis of the microstructure of foams is made more difficult by their low stability. Possible solutions for studying foam structures usually combine optical analytical methods like light microscopy with rheometry. In Fig. 13.10 the results of such combined measurements are presented. The different bubble sizes and corresponding shear stress curves are clearly apparent.

13.10 Chocolate

Without rheology many a Santa might end up an Easter bunny. Like many other products made of chocolate, the chocolate [116] Santa is produced in hollow molds. The pourable chocolate mass is poured into the mold and then centrifuged to distribute it evenly in the mold. Consequently, the flow behavior of the chocolate mass must be adapted 100% to the given processing temperature.

13.11 Psychorheology

The study of texture and its psychological [117] effect is called food psychorheology. Quality is defined primarily by the taste, digestibility and tolerability of food. We perceive the texture of food through the change in the consistency caused by sucking, licking, biting, etc. In the process, our lips, tongue and teeth touch the food. The first bite produces a typical sensation (pudding is soft, chocolate is hard, ...). The force needed for the bite tells us the degree of solidity of the food. The chewing motion provides information on the volume of the food and its resistance. The food mass is diluted with saliva. The final information on the actual texture comes from the throat.

Often eating is only associated with taste. But the appearance of food is also important: would we drink green beer or eat spaghetti if the tomato sauce was blue? Or what if macaroni smelled like pears?

Eating stimulates all our sensory systems:

- Sight (visual)
- Smell (olfactory)
- Touch (tactile)
- Sound (auditory)
- Taste (gustatory)

The overall sensation is called flavor. The appearance of food makes us anticipate the taste, tolerability, digestibility and freshness/doneness. The surface of food can be felt immediately, but the deeper structure is not revealed until food is broken or cut open (e.g. nut).

Another important factor is the temperature of food. Our face is more sensitive than our body to temperature. The lips especially are very thermosensitive. Very cold or hot foods stimulate saliva production for temperature compensation (protection from freezing or burning).

The effect of temperature in the mouth is important for four reasons:

1. Certain foods are expected to elicit a certain behavior (e.g. blowing on soup).
2. The rheological properties depend on the temperature: frozen or fat-containing foods like ice cream or butter change their state from hard to soft/liquid.
3. Odors dissipate as a function of temperature, which is important for flavor perception.

Fig. 13.11. Frost pattern

4. The sensitivity to different taste stimulants varies with the temperature. The highest sensitivity to saccharose is found at 35–50 °C, to salty foods at 18–35 °C and to bitter foods at 10 °C.

Hot/cold sensations can, however, also be elicited by carbonated or alcoholic beverages or by bitter substances. Cooling [118] or warming foods affects our mental state. A hot environment is more likely to promote agitation and physical aggressiveness – a cooling drink can be the antidote.

The texture of food (Fig. 13.11) can also be perceived outside the mouth, for instance when we press a finger against bread dough or judge the ripeness of fruit or a vegetable by its appearance or hear the sound of crushing ice. Fat plays an important role in the evaluation of food consistency for fat is a flavor carrier and therefore we prefer fat-containing food to fat-free food. Fat-containing food (e.g. chocolate) gives a richer mouthfeel, which is assessed positively.

14 List of References

1. M. George: Kleopatra – Der Roman ihres Lebens (two volumes); Verlagsgruppe Lübbe, 2000
2. B. Snell: Heraklitus Ephesius; Fragmente 1944
3. T.X. Thuan: The Birth of the Universe: The Big Bang and After; New York: Harry N. Abrams, 1993
4. H. Kneifel: Die Erde – Planet voller Wunder; Praesentverlag Heinz Peter, Chapter 14, Gütersloh, 1974
5. R.A. Turner: Inventing Leonardo; University of California Press, Berkeley, 1994
6. B. Pascal: Biography in Dictionary of Scientific Biography; Thomson Gale: New York 1970–1990
7. I. Newton: Biography in Encyclopaedia Britannica
8. Chr. Gerthsen; H.O. Kneser: Physik, 11th Edn.; Springer, Berlin Heidelberg New York; 1971, p. 69 and following pgs.
9. R.C. Armstrong; R.B. Bird; O. Hassager: Dynamics of polymeric liquids, 2nd Edn.; Wiley, New York; 1987
10. A. Kolmogorov: Sammelband zur statistischen Theorie der Turbulenz; Akademie-Verlag, Berlin; 1958
11. A. Hermann: Lexikon – Geschichte der Physik A–Z; Aulis-Verlag Deubner & Co., Köln; 1978
12. DIN 53019: Messungen von Viskositäten und Fließkurven mit Rotationsviskosimetern mit Standardgeometrie
13. ASTM: http://www.astm.org/cgi-bin/SoftCart.exe/ABOUT/aboutASTM.html?L+mystore+hwwv3814+1104104891, cited 1 May 2004
14. JSA: http://www.jsa.or.jp/default_english.asp, cited 1 May 2004
15. ISO: http://www.iso.org/iso/en/aboutiso/introduction/index.html, cited 1 May 2004
16. E. Heymann: Haut, Haar und Kosmetik; S. Hirzel Verlag, Stuttgart; 1994
17. K.F. De Polo: A Short Textbook of Cosmetology; Verlag für chemische Industrie, H. Ziolowsky GmbH, Augsburg; 2000
18. R. Brandau: Definitionen der Dermatika unter besonderer Berücksichtigung der Mehrphasensysteme; Arbeitsunterlagen für den Fortbildungskursus "Entwicklung von Emulsionen und Cremes" der Arbeitsgemeinschaft für pharmazeutische Verfahrenstechnik e.V. (APV), Mainz; 1983
19. H. Junginger: Kolloidchemische Betrachtungen an Mehrphasensystemen; Arbeitsunterlagen für den Fortbildungskursus "Entwicklung von Emulsionen und Cremes" der Arbeitsgemeinschaft für pharmazeutische Verfahrenstechnik e.V. (APV), Mainz; 1983
20. Eucerin: Das medizinische Hautpflegeprogramm; Beiersdorf AG Hamburg; 2002
21. T. Tadros: Novel polymeric surfactants in personal care applications – Emulsions, nano-emulsions and multiple emulsions; Sponsored by Orafti NON-Food, Tiene, Belgium; 2002
22. W. Leven: INCI-Index; Synonym-Lexikon der Kosmetikinhaltstoffe, 2nd Edn.; Govi-Verlag, Frankfurt a.M.
23. R. Brummer; T. Berg; W.M. Kulicke: Einfluss des Energieeintrages beim Homogenisieren auf die Struktur von kosmetischen w/o Emulsionen Posterbeitrag; Shaker Verlag, Aachen; 2000
24. H. Armbruster; H. Karbstein; H. Schubert: Chem.-Ing.-Tech. 63(1991), 266
25. D. Exerowa and P.M. Kruglyakov: Foam and Foam Films; Elsevier, Amsterdam; 1998
26. Th.F. Tadros; C. Py; M.C. Taelman; P. Loll: SÖFW 121(10) (1995), 714
27. B. Koglin; J. Pawlowski; H. Schnöring: Chem.-Ing.-Tech. 53(8) (1981), 641
28. K.P. Das; J.E. Kinsella: Advances in Food and Nutrition Research, 34 (1991), 81
29. E. Blaß: Chem.-Ing.-Tech. 60(12) (1988), 935
30. A.K. Chesters: TransIChemE 69A (1991), 259
31. M. Van den Tempel: The Function of Stabilizers during Emulsification. 3rd Int. Congr. Surface Activities, Cologne (2) (1960), 573
32. L. Gehm: Rheologie "Praxisorientierte Grundlagen und Glossar"; Vincentz-Verlag, Hannover; 1998

33. H.-J. Henzler: Rheologische Stoffeigenschaften – Erklärung, Messung, Erfassung und Bedeutung; Chem.-Ing.-Tech. 60(1) (1988), 1
34. H. Schubert: Physikalisch-chemische Grundlagen der Stabilität von o/w Emulsionen; Hochschulkurs Emulgiertechnik Universität Karlsruhe; 1996
35. C. Clasen; W.-M. Kulicke: A convenient way of interpreting steady shear rheo-optical data of semi-dilute polymer solutions, Rheologica Acta 40 (2001), 74–85
36. W. Ostwald: Über die Geschwindigkeitsfunktion der Viskosität disperser Systeme, Kolloid-Zeitschrift 36 (1925), 99–117
37. D. Laba: Rheological Properties of Cosmetics and Toiletries, Marcel Dekker, New York; 1993
38. T. Mezger: Das Rheologie Handbuch; Vincentz-Verlag, Hannover; 2000
39. J.M.G. Cowie: Polymer Chemistry & Physics of Modern Materials, 2nd Edn.; Blackie, Glasgow; 1991
40. P.C. Hiemenz: Polymer Chemistry, The Basic Concepts; Marcel Dekker, New York; 1984
41. M.D. Lechner; K. Gerke; E.H. Nordmeier: Makromolekulare Chemie; Birkhäuser-Verlag, Basel; 1993
42. M. Pahl; W. Gleißle; H.-M. Laun: Praktische Rheologie der Kunststoffe und Elastomere; VDI-Verlag; 1991, p. 66
43. E.B. Bagley: Transactions of the Society of Rheology V pp. 355–368; 1961
44. L.H. Sperling: Introduction to Physical Polymer Science; Wiley, New York; 1986
45. K.P. Menrad: Dynamic Mechanical Analysis; CRC Press, Florida; 1999
46. M. Hediger: Messung rheologischer Eigenschaften; Contraves Industrieprodukte, Stuttgart-Vaihingen; 1965
47. E. Lüscher: Experimentalphysik I, 2. Teil; BI Hochschultaschenbücher, p. 261 and following pgs.
48. E. Kissa: Dispersions-Characterization, Testing and Measurement; Marcel Dekker, New York; 1999
49. B. Schäfer: Lehrbuch der Experimentalphysik, vol. IV, part 1; Aufbau der Materie; Walter de Gruyter, Berlin; 1975, p. 661
50. Report from Bohlin Instruments; 1996
51. Information from GABO Qualimeters; 1993
52. Report from Physica-Meßtechnik; 1999
53. Paper from Haake; 1998
54. Booklet of TA Instruments; 2004
55. News from Göttfert; 2000
56. Information from Brookfield; 2002
57. Booklet of Rheotest Medingen; 2000
58. Booklet of Coesfeld; 2000
59. Marimex: Wir messen Viskosität; Unternehmensbroschüre; 1999
60. F. Durst; A. Melling; J.H. Whitelaw: Principles and practice of Laser-Doppler Anemometry; Academic, London; 1976
61. D.Q.M. Craig: Dielectric Analysis of Pharmaceutical Systems; Taylor & Francis, London; 1995
62. J. Haag: Praktische Rheologie – Bestimmung der Fließgrenze; Firmenschrift 2/92 der Firma Bohlin Instruments, Mühlacker; 1992
63. R. Brummer; G. Hamer: Rheological Methods to Characterize Cosmetic Products; Second World Congress on Emulsion, vol. 2, Theme: 2-4/p. 128; Bordeaux; 1997
64. T. Metzger: Lohnt sich Rheologie? Physica Messtechnik GmbH, Stuttgart; 2001
65. F. Hetzel; J. Nielsen; S. Wiesner; R. Brummer: Dynamic mechanical freezing points of cosmetic o/w emulsions and their stability at low temperatures; Applied Rheology 10(3) (2000), 114–118
66. R. Brummer; M. Griebenow; F. Hetzel; V. Schlesiger; R. Uhlmann: Rheological Swing Test to Predict the Temperature Stability of Cosmetic Emulsions; Verlag für chemische Industrie, H. Ziolkowsky GmbH, Augsburg, Germany; Proceedings XXI IFSCC International Congress 2000, Berlin; pp. 476
67. P. Stern: Die Rheologie in der Kosmetik; Zweites Rheologiesymposium der DDR, Vortragsband 2, Tabarz, Germany; 1987
68. R. Brummer; S. Godersky: Rheological studies to objectify sensations occurring when cosmetic emulsions are applied to the skin: Colloids and Surfaces A – Physiochemical and Engineering Aspects 1999 152(1,2) (1999), 89
69. F. Hetzel; P. Gluth; R. Brummer; K.-P. Wittern: Rheological determination of the pumping and emptying behaviour of cosmetic emulsions, 6th European Conference on Rheology; Erlangen; 2002

70. R. Brummer; T. Berg; W.-M. Kulicke: Einfluss des Energieeintrages beim Homogenisieren auf die Struktur von kosmetischen w/o Emulsionen Posterbeitrag, Shaker-Verlag, Aachen; 2002
71. T. Berg; P. Arlt; R. Brummer; D. Emeis; W.-M. Kulicke; S. Wiesner; K.-P. Wittern: Insights into the structure and dynamics of complex w/o-emulsions by combining NMR, rheology and electron microscopy; Colloids and Surfaces A. Physcochem. Eng. Aspects 238 (2004), 59–59
72. U.T. Lashmar; J. Beesley: Int. J. Pharm. 91 (1993), 59
73. Malvern Instruments: www.mastersizer.com; 1999
74. M. Heuer; K. Leschonski: 3rd Europäisches Symposium Partikelmeßtechnik, Nürnberg; 1984, p. 516
75. G. De Boer: Part. Syst. Charact. 4 (1987), 14
76. T. Allen: Particle Size Measurements, Chapman and Hall, London; 1990
77. H. Schuchmann: Hochschulkurs Emulgiertechnik, Universität Karlsruhe; 1994
78. S. Schaugsdat: Rheologische Untersuchung zur Stabilität kosmetischer Emulsionen; Diplomarbeit an der Fachhochschule Hamburg; 1995
79. W. Gleissle: Rheologica Acta, 21 (1982), 484
80. W.-M. Kulicke: Fließverhalten von Stoffen und Stoffgemischen, Hüthig & Wepf Verlag, Basel; 1986
81. J.D. Ferry: Viscoelastic Properties of Polymers; Wiley & Sons, New York; 1970
82. S. Arrhenius: On the Reaction Velocity of the Inversion of Cane Sugar by Acids; Zeitschrift für physikalische Chemie 4 (1889) p. 226 and following pgs.
83. M.L. Williams; R.F. Landel; J.D. Ferry: Journal of the American Chemical Society 77 (1955), 3701
84. M. Mühlbach; R. Brummer; R. Eggers: Überprüfung der Anwendung des Time Temperature Superposition-Prinzips auf Emulsionen; Journal of Cosmetic Science, 2005
85. W.P. Cox; E.H. Merz: Correlation of Dynamic and Steady Flow Viscosities. Journal of Polymer Science 28(118) (1958), 619
86. W.-M. Kulicke; R.S. Porter: Relation between steady shear flow and dynamic rheology; Rheologica Acta, 19 (1980), 601
87. M.L. Huggins: J. Am. Chem. Soc. 64 (1942), 623
88. T. Berg: Diplomarbeit: Chemische und sterische Strukturbestimmung von Naturkautschuken; Institut für Technische und Makromolekulare Chemie der Universität Hamburg; 1996
89. W. Pfandl; F.R. Schwarzl: Relation between Rheological Properties and the Molecular Weight Distribution of a Technical Polystyrene, Colloid & Polym. Sci. 263, (1985), 328
90. R. Brummer: Das Rheometer der Zukunft, Vortrag auf dem RheoFuture Meeting bei Haake, Karlsruhe; 2001
91. Standard Meter: Article taken from “Backsights” Magazine published by Surveyors Historical Society, Internet
92. H. Viskosimeter: Einführung in die praktische Viskosimetrie; Gebrüder Haake GmbH; 1981, p. 7
93. L. Sachs: Angewandte Statistik; Springer, Berlin Heidelberg New York; 2000
94. B. Eike: Statistik – Eine Einführung; Pythagoras Lehrmittel, 1st Edn.; 2003, p. 103 and following pgs.
95. I. Bronstein; K. Semendjajew: Taschenbuch der Mathematik, part VI; Verlag Harri Deutsch, Zurich; 1974
96. H. Schmidt: Was ist Genauigkeit? Zum Einfluß systematischer Abweichungen auf Meß- und Ausgleichungsergebnisse; Geodätisches Institut der RWTH Aachen; 1999
97. EuroPhysics: Rheologie und Rheometrie für Anwender; EuroPhysics Fachbroschüre Rheologie; 1998
98. R. Brummer; C.H. Walther: Kosmetische Emulsionen rheologisch richtig messen; Parfümerie und Kosmetik 79(1008) 01-02, 16
99. W. Umbach: Kosmetik – Entwicklung, Herstellung und Anwendung kosmetischer Mittel. Thieme-Verlag, Stuttgart; 1995
100. Under § 5 d Para. 1 of the Cosmetic Directive
101. § 34 Para. 1 MBO (Model Professional Regulation for German Doctors)
102. Law Prohibiting Unfair Competition (UWG)
103. Food and Commodities Law (LBMG)
104. Drug Advertising Law (HWG)
105. D. Weipert; H.-D. Tscheuschner; E. Windhab: Rheologie der Lebensmittel; Behrs-Verlag, Hamburg; 1993
106. Brockhaus Enzyklopädie, 19th Edn., vol. 24, vol. 15; F.A. Brockhaus, Mannheim; 1991
107. Ph. Born: Der Konditor der Neuzeit; B.F. Voigt, Leipzig; 1912
108. O. Rahn; P.F. Sharp: Physik der Milchwirtschaft; P. Parey, Berlin; 1928
109. J. Pryce-Jones: The rheology of honey. In: G.W.S. Blair (ed.): Foodstuff – their Plasticy, Fluidity and Consistency; Amsterdam, North Holland; 1953

110. Handbook: Rheology and food technology; Rheometrics, Pisscataway, NJ; 1987
111. M. Baron; G.W.S. Blair: Cheese and Curd. In: G.W.S. Blair (ed.): Foodstuff – their Plasticy, Fluidity and Consistency. North Holland, Amsterdam; 1953
112. T. Mezger: Das Rheologie Handbuch; Vincentz-Verlag; 2000, p. 56
113. K. Schulze; H.-J. Räuber; K. Guthy: Viskoelastische Eigenschaften der Milch während der Joghurtfermentierung, Rheology 91(10) (1991), 160
114. C. Pieske: Marzipan aus Lübeck; Weiland G. Nachf., Lübeck; 1997
115. C. Rha (ed.): Theory, determination and control of physical properties of food materials; Reidel, Dordrecht; 1975
116. H.D. Tscheuchner; D. Wünsche: Rheological properties of chocolate masses and the influence of some factors. In P. Sherman (ed.): Food Texture and Rheology. Academic, London; 1979
117. M.C. Bourne: Food Texture and Viscosity: Concept and Measurement; Academic Press, Washington; 2002
118. T. Metzger: Vanilleeis im Rheometer; Analytica Pro, Frankfurt a.M.; 2004, p. 56

15 Subject Index

Printing: Krips bv, Meppel
Binding: Stürtz, Würzburg